STUDENT SOLUTIONS MANUAL

CHAPTERS 20–43

PHYSICS

FOR SCIENTISTS AND ENGINEERS SECOND EDITION

A STRATEGIC APPROACH

Randall D. Knight

Pawan Kahol
Missouri State University

Donald Foster
Wichita State University

Larry Smith
Snow College

Scott Nutter
Northern Kentucky University

San Francisco Boston New York
Cape Town Hong Kong London Madrid Mexico City
Montreal Munich Paris Singapore Sydney Tokyo Toronto

Publisher: Adam Black, Ph.D.
Development Manager: Michael Gillespie
Development Editor: Alice Houston, Ph.D.
Project Editor: Martha Steele
Assistant Editor: Grace Joo
Media Producer: Deb Greco
Sr. Administrative Assistant: Cathy Glenn
Director of Marketing: Christy Lawrence
Executive Marketing Manager: Scott Dustan
Sr. Market Development Manager: Josh Frost
Market Development Associate: Jessica Lyons
Managing Editor: Corinne Benson
Sr. Production Supervisor: Nancy Tabor
Production Service: WestWords PMG
Illustrations: International Typesetting and Composition (ITC)
Cover Design: Yvo Riezebos Design and Seventeenth Street Studios
Manufacturing Manager: Evelyn Beaton
Manufacturing Buyer: Carol Melville
Text and Cover Printer: Bradford & Bigelow
Cover Image: Composite illustration by Yvo Riezebos Design; photo of spring by Bill Frymire/Masterfile

ISBN-13: 978-0-321-51356-4
ISBN-10: 0-321-51356-8

www.aw-bc.com

3 4 5 6 7 8 9 10—B&B—12 11 10 09

Contents

Preface

This *Student Solutions Manual* is intended to provide you with examples of good problem-solving techniques and strategies. To achieve that, the solutions presented here attempt to:.

- Follow, in detail, the problem-solving strategies presented in the text.
- Articulate the reasoning that must be done before computation.
- Illustrate how to use drawings effectively.
- Demonstrate how to utilize graphs, ratios, units, and the many other "tactics" that must be successfully mastered and marshaled if a problem-solving strategy is to be effective.
- Show examples of assessing the reasonableness of a solution.
- Comment on the significance of a solution or on its relationship to other problems.

We recommend you try to solve each problem on your own before you read the solution. Simply reading solutions, without first struggling with the issues, has limited educational value.

As you work through each solution, make sure you understand how and why each step is taken. See if you can understand which aspects of the problem made this solution strategy appropriate. You will be successful on exams not by memorizing solutions to particular problems but by coming to recognize which kinds of problem-solving strategies go with which types of problems.

We have made every effort to be accurate and correct in these solutions. However, if you do find errors or ambiguities, we would be very grateful to hear from you. Please contact: knight@aw.com

Acknowledgments for the First Edition

We are grateful for many helpful comments from Susan Cable, Randall Knight, and Steve Stonebraker. We express appreciation to Susan Emerson, who typed the word-processing manuscript, for her diligence in interpreting our handwritten copy. Finally, we would like to acknowledge the support from the Addison Wesley staff in getting the work into a publishable state. Our special thanks to Liana Allday, Alice Houston, and Sue Kimber for their willingness and preparedness in providing needed help at all times.

Pawan Kahol
Missouri State University

Donald Foster
Wichita State University

Acknowledgments for the Second Edition

I would like to acknowledge the patient support of my wife, Holly, who knows what is important.

Larry Smith
Snow College

I would like to acknowledge the assistance and support of my wife, Alice Nutter, who helped type many problems and was patient while I worked weekends.

Scott Nutter
Northern Kentucky University

20

TRAVELING WAVES

20.3. Model: The wave pulse is a traveling wave on a stretched string.
Solve: The wave speed on a stretched string with linear density μ is

$$v_{\text{string}} = \sqrt{\frac{T_S}{\mu}} = \sqrt{\frac{T_S}{m/L}} = \sqrt{\frac{LT_S}{m}} \Rightarrow \frac{2.0 \text{ m}}{50 \times 10^{-3} \text{ s}} = \sqrt{\frac{(2.0 \text{ m})(20 \text{ N})}{m}} \Rightarrow m = 0.025 \text{ kg} = 25 \text{ g}$$

20.5. Model: This is a wave traveling at constant speed. The pulse moves 1.0 m to the left every second.
Visualize: This snapshot graph shows the wave at all points on the x-axis at $t = 2.0$ s. You can see that the leading edge of the wave at $t = 1.0$ s is precisely at $x = 0.0$ m. That is, in the first second, the displacement is zero at $x = 0.0$ m. The first part of the wave causes a upward displacement of the medium, so immediately after $t = 1.0$ s the displacement at $x = 0.0$ m will be positive. The positive portion of the wave pulse is 2.0 m wide and takes 2.0 s to pass $x = 0.0$ m. The negative portion begins to pass through $x = 0.0$ m at $t = 3.0$ s and until $t = 5.0$ s the displacement of the medium is negative. The displacement at $x = 0.0$ m returns to zero at $t = 5.0$ s and remains zero for all later times.

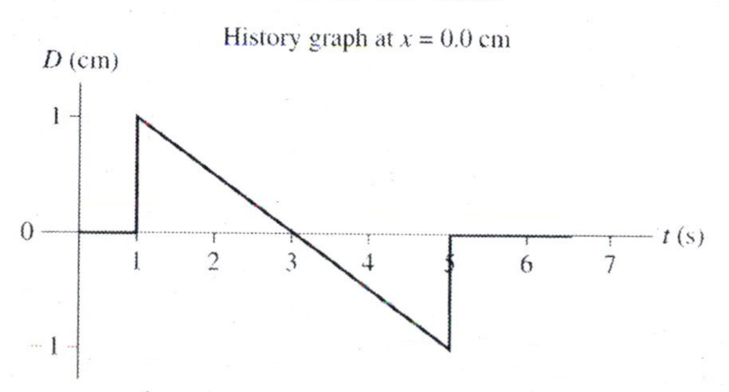

20.7. Model: This is a wave traveling at constant speed to the left at 1.0 m/s.
Visualize: This is the history graph of a wave at $x = 2.0$ m. Because the wave is moving to the left at 1.0 m/s, the wave passes the $x = 2.0$ m position a distance of 1.0 m in 1.0 s. Because the width of the linearly increasing part of the history graph takes 2.0 s to pass the $x = 2.0$ m position, its width is 2.0 m. Similarly, the width of the flat part of the history graph is 2.0m.

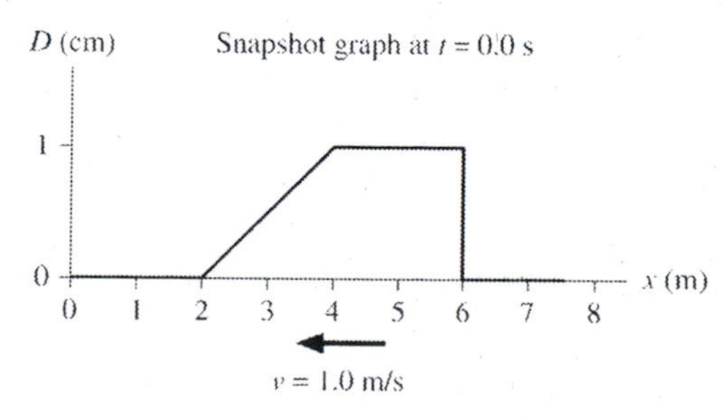

20.9. Visualize:

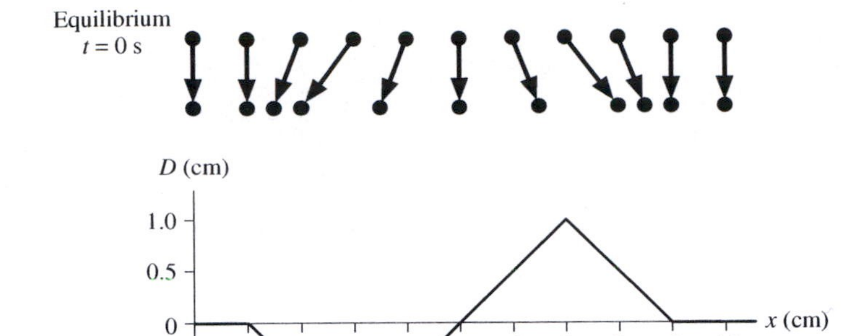

We first draw the particles of the medium in the equilibrium positions, with an inter-particle spacing of 1.0 cm. Just underneath, the positions of the particles as a longitudinal wave is passing through are shown at time $t = 0$ s. It is clear that relative to the equilibrium the particle positions are displaced negatively on the left side and positively on the right side. For example, the particles at $x = 0$ cm and $x = 1$ cm are at equilibrium, the particle at $x = 2$ cm is displaced left by 0.5 cm, the particle at $x = 3$ cm is displaced left by 1.0 cm, the particle at $x = 4$ cm is displaced left by 0.5 cm, and the particle at $x = 5$ cm is undisplaced. The behavior of particles for $x > 5$ cm is opposite of that for $x < 5$ cm.

20.13. Model: The wave is a traveling wave.

Solve: **(a)** A comparison of the wave equation with Equation 20.14 yields: $A = 5.2$ cm, $k = 5.5$ rad/m, $\omega = 72$ rad/s, and $\phi_0 = 0$ rad. The frequency is

$$f = \frac{\omega}{2\pi} = \frac{72 \text{ rad/s}}{2\pi} = 11.5 \text{ Hz} \approx 11 \text{ Hz}$$

(b) The wavelength is

$$\lambda = \frac{2\pi}{k} = \frac{2\pi}{5.5 \text{ rad/m}} = 1.14 \text{ m} \approx 1.1 \text{ m}$$

(c) The wave speed $v = \lambda f = 13$ m/s.

20.17. Visualize:

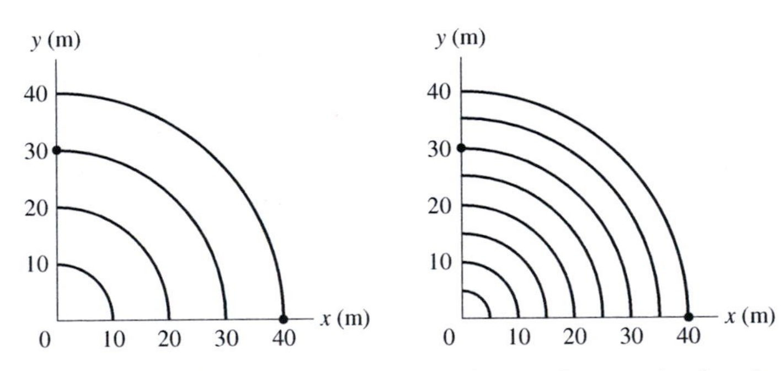

Solve: For a sinusoidal wave, the phase difference between two points on the wave is given by Equation 20.28:

$$\Delta\phi = \phi_2 - \phi_1 = \frac{2\pi}{\lambda}(r_2 - r_1) = \frac{2\pi}{\lambda}(40 \text{ m} - 30 \text{ m}) \Rightarrow \lambda = \frac{2\pi}{\Delta\phi}(10 \text{ m})$$

$\Delta\phi = 2\pi$ for two points on adjacent wavefronts and $\Delta\phi = 4\pi$ for two points separated by 2λ. Thus, λ = 10 m when $\Delta\phi = 2\pi$, and λ = 5 m when $\Delta\phi = 4\pi$. The crests corresponding to these two wavelengths are shown in the figure. One can see that a crest of the wave passes the 40 m–listener and the 30 m–listener simultaneously. The lowest two possible frequencies will occur for the largest two possible wavelengths, which are 10 m and 5 m. Thus, the lowest frequency is

$$f_1 = \frac{v}{\lambda} = \frac{340 \text{ m/s}}{10 \text{ m}} = 34 \text{ Hz}$$

The next highest frequency is $f_2 = 68$ Hz.

20.25. Model: Light is an electromagnetic wave.
Solve: **(a)** The time light takes is

$$t = \frac{3.0 \text{ mm}}{v_{\text{glass}}} = \frac{3.0\times10^{-3} \text{ m}}{c/n} = \frac{3.0\times10^{-3} \text{ m}}{\left(3.0\times10^{8} \text{ m/s}\right)/1.50} = 1.5\times10^{-11} \text{ s}$$

(b) The thickness of water is

$$d = v_{\text{water}}t = \frac{c}{n_{\text{water}}}t = \frac{3.0\times10^{8} \text{ m/s}}{1.33}\left(1.5\times10^{-11} \text{ s}\right) = 3.4 \text{ mm}$$

20.31. Solve: **(a)** The intensity of a uniform spherical source of power P_{source} a distance r away is $I = P_{\text{source}}/4\pi r^2$. Thus, the intensity at the position of the microphone is

$$I_{50 \text{ m}} = \frac{35 \text{ W}}{4\pi(50 \text{ m})^2} = 1.1\times10^{-3} \text{ W/m}^2$$

(b) The sound energy impinging on the microphone per second is

$$P = Ia = \left(1.1\times10^{-3} \text{ W/m}^2\right)\left(1.0\times10^{-4} \text{ m}^2\right) = 1.1\times10^{-7} \text{ W} = 1.1\times10^{-7} \text{ J/s}$$

$$\Rightarrow \text{Energy impinging on the microphone in 1 second} = 1.1\times10^{-7} \text{ J}$$

20.37. Model: Your friend's frequency is altered by the Doppler effect. The frequency of your friend's note increases as he races towards you (moving source and a stationary observer). The frequency of your note for your approaching friend is also higher (stationary source and a moving observer).
Solve: **(a)** The frequency of your friend's note as heard by you is

$$f_+ = \frac{f_0}{1-\dfrac{v_S}{v}} = \frac{400 \text{ Hz}}{1-\dfrac{25.0 \text{ m/s}}{340 \text{ m/s}}} = 432 \text{ Hz}$$

(b) The frequency heard by your friend of your note is

$$f_+ = f_0\left(1+\frac{v_0}{v}\right) = (400 \text{ Hz})\left(1+\frac{25.0 \text{ m/s}}{340 \text{ m/s}}\right) = 429 \text{ Hz}$$

20.45. Model: The wave pulse is a traveling wave on a stretched string.
Solve: While the tension T_S is the same in both the strings, the wave speeds in the two strings are not. We have

$$v_1 = \sqrt{\frac{T_S}{\mu_1}} \quad \text{and} \quad v_2 = \sqrt{\frac{T_S}{\mu_2}} \Rightarrow v_1^2\mu_1 = v_2^2\mu_2 = T_S$$

Because $v_1 = L_1/t_1$ and $v_2 = L_2/t_2$, and because the pulses are to reach the ends of the string simultaneously, the above equation can be simplified to

$$\frac{L_1^2\mu_1}{t^2} = \frac{L_2^2\mu_2}{t^2} \Rightarrow \frac{L_1}{L_2} = \sqrt{\frac{\mu_2}{\mu_1}} = \sqrt{\frac{4.0 \text{ g/m}}{2.0 \text{ g/m}}} = \sqrt{2} \Rightarrow L_1 = \sqrt{2}L_2$$

Since $L_1 + L_2 = 4$ m,

$$\sqrt{2}L_2 + L_2 = 4\text{ m} \Rightarrow L_2 = 1.66\text{ m} \approx 1.7\text{ m} \quad \text{and} \quad L_1 = \sqrt{2}(1.66\text{ m}) = 2.34\text{ m} \approx 2.3\text{ m}$$

20.47. Visualize:

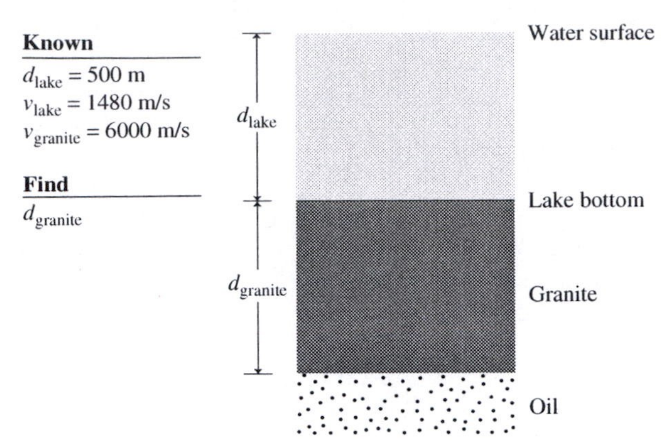

Solve: The explosive's sound travels down the lake and into the granite, and then it is reflected by the oil surface. The echo time is thus equal to

$$t_{echo} = t_{water\ down} + t_{granite\ down} + t_{granite\ up} + t_{water\ up}$$

$$0.94\text{ s} = \frac{500\text{ m}}{1480\text{ m/s}} + \frac{d_{granite}}{6000\text{ m/s}} + \frac{d_{granite}}{6000\text{ m/s}} + \frac{500\text{ m}}{1480\text{ m/s}} \Rightarrow d_{granite} = 790\text{ m}$$

20.53. Model: This is a sinusoidal wave.
Solve: **(a)** The displacement of a wave traveling in the positive x-direction with wave speed v must be of the form $D(x, t) = D(x - vt)$. Since the variables x and t in the given wave equation appear together as $x + vt$, the wave is traveling toward the left, that is, in the $-x$ direction.
(b) The speed of the wave is

$$v = \frac{\omega}{k} = \frac{2\pi/0.20\text{ s}}{2\pi\text{ rad}/2.4\text{ m}} = 12\text{ m/s}$$

The frequency is

$$f = \frac{\omega}{2\pi} = \frac{2\pi\text{ rad}/0.20\text{ s}}{2\pi} = 5.0\text{ Hz}$$

The wave number is

$$k = \frac{2\pi\text{ rad}}{2.4\text{ m}} = 2.6\text{ rad/m}$$

(c) The displacement is

$$D(0.20\text{ m}, 0.50\text{ s}) = (3.0\text{ cm})\sin\left[2\pi\left(\frac{0.20\text{ m}}{2.4\text{ m}} + \frac{0.50\text{ s}}{0.20\text{ s}} + 1\right)\right] = -1.5\text{ cm}$$

20.61. Model: This is a wave traveling to the left at a constant speed of 50 cm/s.
Solve: The particles at positions between $x = 2$ cm and $x = 7$ cm have a speed of 10 cm/s, and the particles between $x = 7$ cm and $x = 9$ cm have a speed of -25 cm/s. That is, at the time the snapshot of the velocity is shown, the particles of the medium have upward motion for $2\text{ cm} \le x \le 7\text{ cm}$, but downward motion for $7\text{ cm} \le x \le 9\text{ cm}$. The width of the front section of the wave pulse is $7\text{ cm} - 2\text{ cm} = 5$ cm and the width of the rear section is $9\text{ cm} - 7\text{ cm} = 2$ cm. With a wave speed of 50 cm/s, the time taken by the front section to pass through a particular point is $5\text{ cm}/50\text{ cm/s} = 0.1\text{ s}$ and the time taken by the rear section of the wave to pass through a point is $2\text{ cm}/50\text{ cm/s} = 0.04\text{ s}$. Thus the wave causes the upward moving particles to go through a

displacement of $A = (10\text{ cm/s})(0.1\text{ s}) = 1.0\text{ cm}$. The downward moving particles have a maximum displacement of $(-25\text{ cm/s})(0.04\text{ s}) = -1.0\text{ cm}$.

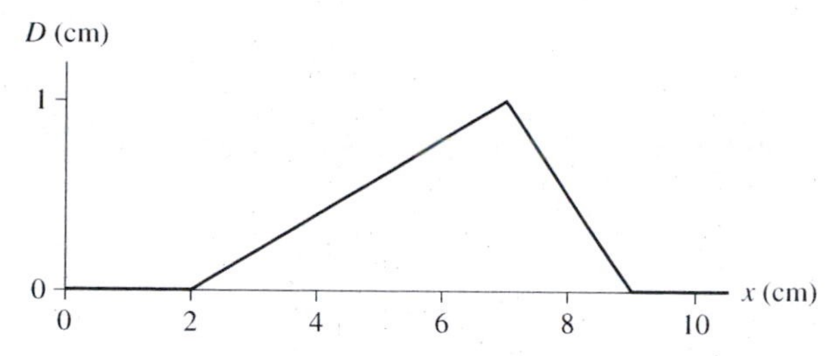

20.67. Model: We have a traveling wave radiated by the tornado siren.
Solve: **(a)** The power of the source is calculated as follows:

$$I_{50\text{ m}} = 0.10\text{ W/m}^2 = \frac{P_{\text{source}}}{4\pi r^2} = \frac{P_{\text{source}}}{4\pi(50\text{ m})^2} \Rightarrow P_{\text{source}} = (0.10\text{ W/m}^2)4\pi(50\text{ m})^2 = (1000\pi)\text{ W}$$

The intensity at 1000 m is

$$I_{1000\text{ m}} = \frac{P_{\text{source}}}{4\pi(1000\text{ m})^2} = \frac{(1000\pi)\text{ W}}{4\pi(1000\text{ m})^2} = 250\ \mu\text{W/m}^2$$

(b) The maximum distance is calculated as follows:

$$I = \frac{P_{\text{source}}}{4\pi r^2} \Rightarrow 1.0\times10^{-6}\text{ W/m}^2 = \frac{(1000\pi)\text{ W}}{4\pi r^2} \Rightarrow r = 16\text{ km}$$

20.69. Model: Assume the two loudspeakers broadcast the same power and that the platforms are high enough off the ground that we don't have to worry about reflected sound.
Visualize: Call the distance between the loudspeakers d. Call the intensity halfway between the speakers (at $d/2$) I_1 and the sound intensity level there $\beta_1(=75dB)$; call them I_2 and β_2 at 1/4 the distance from one pole and 3/4 the distance from the other pole on the line between them. We seek β_2.
We first apply a general approach for different sound intensity levels:

$$\Delta\beta = \beta_2 - \beta_1 = (10\text{dB})\left[\log_{10}\left(\frac{I_2}{I_0}\right) - \log_{10}\left(\frac{I_1}{I_0}\right)\right] = (10\text{ dB})\log_{10}\left(\frac{I_2/I_0}{I_1/I_0}\right) = (10\text{ dB})\log_{10}\left(\frac{I_2}{I_1}\right)$$

Solve: Recall that for the general case of spherical symmetry $I = P/A$, where P is the power emitted by the source and $A = 4\pi R^2$ is the area of the sphere. Now we find the ratio of the intensities I_2/I_1 and then plug it in the formula above and add it to 75 dB.

$$I_1 = \frac{P}{4\pi(d/2)^2} + \frac{P}{4\pi(d/2)^2} = \frac{2P}{\pi d^2}$$

$$I_2 = \frac{P}{4\pi(d/4)^2} + \frac{P}{4\pi(3d/4)^2} = \frac{4P}{\pi d^2} + \frac{4P}{9\pi d^2} = \frac{(36+4)P}{9\pi d^2} = \frac{40P}{9\pi d^2} = \frac{20}{9}I_1$$

$$\Delta\beta = (10\text{ dB})\log_{10}\left(\frac{I_2}{I_1}\right) = (10\text{ dB})\log_{10}\left(\frac{20}{9}\right) = 3.48\text{ dB}$$

$$\beta_2 = \beta_1 + \Delta\beta = 75\text{ dB} + 3.48\text{ dB} = 78\text{ dB}$$

Assess: An increase of about 3dB corresponds to a doubling of the intensity. 20/9 is close to double.

20.75. Model: We are looking at the Doppler effect for the light of a receding source.
Visualize:

You
$0.1c$
Light
Daredevil
$0.3c$
Jupiter

Note that the daredevil's tail lights are receding away from your rocket's light detector with a relative speed of $0.2c$.
Solve: Using Equation 20.40, the observed wavelength is

$$\lambda = \sqrt{\frac{1+v_s/c}{1-v_s/c}}\lambda_0 = \sqrt{\frac{1+0.2c/c}{1-0.2c/c}}(650\text{ nm}) = 796\text{ nm} \approx 800\text{ nm}$$

This wavelength is in the infrared region.

20.77. Model: The Doppler effect for light of an approaching source leads to a decreased wavelength.
Solve: The red wavelength ($\lambda_0 = 650$ nm) is Doppler shifted to green ($\lambda = 540$ nm) due to the approaching light source. In relativity theory, the distinction between the motion of the source and the motion of the observer disappears. What matters is the relative approaching or receding motion between the source and the observer. Thus, we can use Equation 20.40 as follows:

$$\lambda = \lambda_0\sqrt{\frac{1-v_s/c}{1+v_s/c}} \Rightarrow 540\text{ nm} = (650\text{ nm})\sqrt{\frac{1-v_s/c}{1+v_s/c}}$$

$$\Rightarrow v_s = 5.5\times10^4\text{ km/s} = 2.0\times10^8\text{ km/h}$$

The fine will be

$$\left(2.0\times10^8\text{ km/hr} - 50\text{ km/hr}\right)\left(\frac{1\,\$}{1\text{ km/hr}}\right) = \$200\text{ million}$$

Assess: The police officer knew his physics.

21

SUPERPOSITION

Exercises and Problems

21.3. Model: The principle of superposition comes into play whenever the waves overlap.
Visualize:

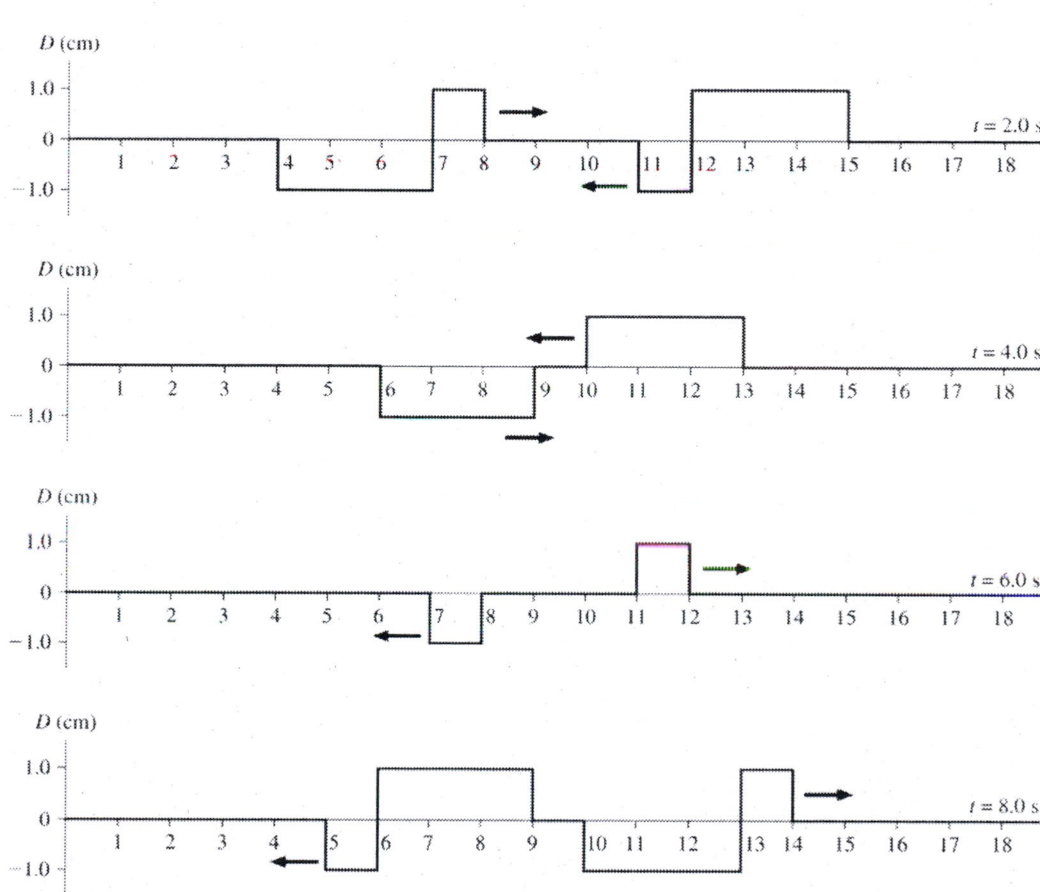

At $t = 4.0$ s the shorter pulses overlap and cancel. At $t = 6.0$ s the longer pulses overlap and cancel.

21.9. Model: A string fixed at both ends supports standing waves.
Solve: (a) We have $f_a = 24\text{ Hz} = mf_1$ where f_1 is the fundamental frequency that corresponds to $m = 1$. The next successive frequency is $f_b = 36\text{ Hz} = (m+1)f_1$. Thus,

$$\frac{f_b}{f_a} = \frac{(m+1)f_1}{mf_1} = \frac{m+1}{m} = \frac{36\text{ Hz}}{24\text{ Hz}} = 1.5 \Rightarrow m+1 = 1.5m \Rightarrow m = 2 \Rightarrow f_1 = \frac{24\text{ Hz}}{2} = 12\text{ Hz}$$

The wave speed is

$$v = \lambda_1 f_1 = \frac{2L}{1} f_1 = (2.0\ \text{m})(12\ \text{Hz}) = 24\ \text{m/s}$$

(b) The frequency of the third harmonic is 36 Hz. For $m = 3$, the wavelength is

$$\lambda_m = \frac{2L}{m} = \frac{2(1\ \text{m})}{3} = \frac{2}{3}\ \text{m}$$

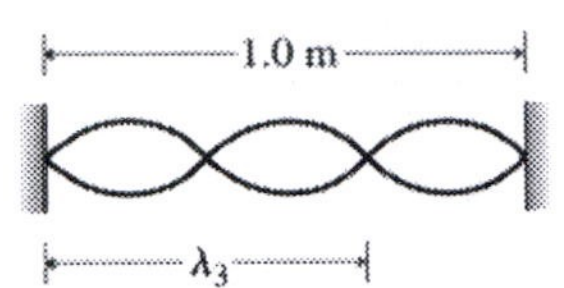

21.17. Model: An organ pipe has a "sounding" hole where compressed air is blown across the edge of the pipe. This is one end of an open-open tube with the other end at the true "end" of the pipe.
Solve: For an open-open tube, the fundamental frequency is $f_1 = 16.4$ Hz. We have

$$\lambda_1 = \frac{2L}{1} \Rightarrow L = \frac{\lambda_1}{2} = \frac{1}{2}\left(\frac{v_{\text{sound}}}{f_1}\right) = \frac{1}{2}\left(\frac{343\ \text{m/s}}{16.4\ \text{Hz}}\right) = 10.5\ \text{m}$$

Assess: The length of the organ pipe is ≈ 34.5 feet. That is actually somewhat of an overestimate since the antinodes of real tubes are slightly outside the tube. The actual length in a real organ is about 32 feet, and this is the tallest pipe in the so called "32 foot rank" of pipes.

21.19. Model: A string fixed at both ends forms standing waves.
Solve: A simple string sounds the fundamental frequency $f_1 = v/2L$. Initially, when the string is of length $L_A = 30$ cm, the note has the frequency $f_{1A} = v/2L_A$. For a different length, $f_{1B} = v/2L_B$. Taking the ratio of each side of these two equations gives

$$\frac{f_{1A}}{f_{1B}} = \frac{v/2L_A}{v/2L_B} = \frac{L_B}{L_A} \Rightarrow L_B = \frac{f_{1A}}{f_{1B}} L_A$$

We know that the second frequency is desired to be $f_{1B} = 523$ Hz. The string length must be

$$L_B = \frac{440\ \text{Hz}}{523\ \text{Hz}}(30\ \text{cm}) = 25.2\ \text{cm}$$

The question is not how long the string must be, but where must the violinist place his finger. The full string is 30 cm long, so the violinist must place his finger 4.8 cm from the end.
Assess: A fingering distance of 4.8 cm from the end is reasonable.

21.21. Model: The interference of two waves depends on the difference between the phases $(\Delta\phi)$ of the two waves.
Solve: **(a)** Because the speakers are in phase, $\Delta\phi_0 = 0$ rad. Let d represent the path-length difference. Using $m = 0$ for the smallest d and the condition for destructive interference, we get

$$\Delta\phi = 2\pi\frac{\Delta x}{\lambda} + \Delta\phi_0 = 2\left(m + \tfrac{1}{2}\right)\pi\ \text{rad} \qquad m = 0, 1, 2, 3 \ldots$$

$$\Rightarrow 2\pi\frac{\Delta x}{\lambda} + \Delta\phi_0 = \pi\ \text{rad} \Rightarrow 2\pi\frac{d}{\lambda} + 0\ \text{rad} = \pi\ \text{rad} \Rightarrow d = \frac{\lambda}{2} = \frac{1}{2}\left(\frac{v}{f}\right) = \frac{1}{2}\left(\frac{343\ \text{m/s}}{686\ \text{Hz}}\right) = 0.25\ \text{m}$$

(b) When the speakers are out of phase, $\Delta\phi_0 = \pi$. Using $m = 1$ for the smallest d and the condition for constructive interference, we get

$$\Delta\phi = 2\pi\frac{\Delta x}{\lambda} + \Delta\phi_0 = 2m\pi \qquad m = 0, 1, 2, 3, \ldots$$

$$\Rightarrow 2\pi\frac{d}{\lambda} + \pi = 2\pi \Rightarrow d = \frac{\lambda}{2} = \frac{1}{2}\left(\frac{v}{f}\right) = \frac{1}{2}\left(\frac{343\ \text{m/s}}{686\ \text{Hz}}\right) = 0.25\ \text{m}$$

21.23. Model: Reflection is maximized if the two reflected waves interfere constructively.
Solve: The film thickness that causes constructive interference at wavelength λ is given by Equation 21.32:

$$\lambda_C = \frac{2nd}{m} \Rightarrow d = \frac{\lambda_C m}{2n} = \frac{(600\times10^{-9}\text{ m})(1)}{(2)(1.39)} = 216\text{ nm}$$

where we have used $m = 1$ to calculate the thinnest film.
Assess: The film thickness is much less than the wavelength of visible light. The above formula is applicable because $n_{\text{air}} < n_{\text{film}} < n_{\text{glass}}$.

21.27. Model: The two speakers are identical, and so they are emitting circular waves in phase. The overlap of these waves causes interference.
Visualize:

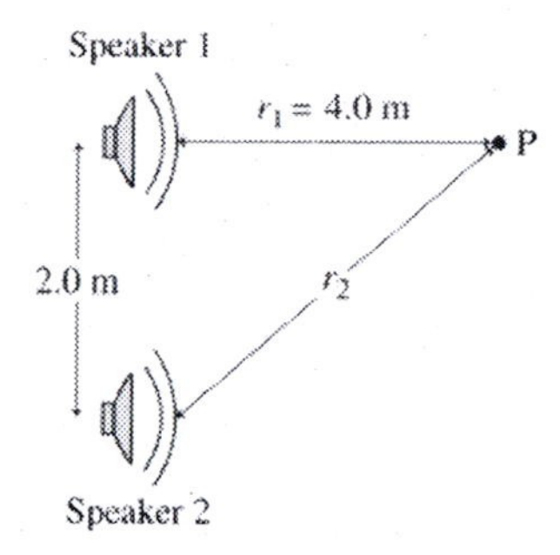

Solve: From the geometry of the figure,

$$r_2 = \sqrt{r_1^2 + (2.0\text{ m})^2} = \sqrt{(4.0\text{ m})^2 + (2.0\text{ m})^2} = 4.472\text{ m}$$

So, $\Delta r = r_2 - r_1 = 4.472\text{ m} - 4.0\text{ m} = 0.472\text{ m}$. The phase difference between the sources is $\Delta\phi_0 = 0\text{ rad}$ and the wavelength of the sound waves is

$$\lambda = \frac{v}{f} = \frac{340\text{ m/s}}{1800\text{ Hz}} = 0.1889\text{ m}$$

Thus, the phase difference of the waves at the point 4.0 m in front of one source is

$$\Delta\phi = 2\pi\frac{\Delta r}{\lambda} + \Delta\phi_0 = \frac{2\pi(0.472\text{ m})}{0.1889\text{ m}} + 0\text{ rad} = 5\pi\text{ rad} = 2.5(2\pi\text{ rad})$$

This is a half-integer multiple of 2π rad, so the interference is perfect destructive.

21.29. Solve: The beat frequency is

$$f_{\text{beat}} = f_1 - f_2 \Rightarrow 3\text{ Hz} = f_1 - 200\text{ Hz} \Rightarrow f_1 = 203\text{ Hz}$$

f_1 is larger than f_2 because the increased tension increases the wave speed and hence the frequency.

21.33. Model: The wavelength of the standing wave on a string vibrating at its second-harmonic frequency is equal to the string's length.
Visualize:

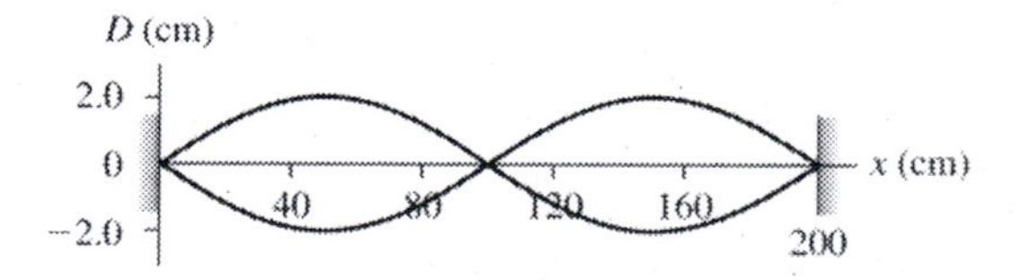

Solve: The length of the string $L = 2.0$ m, so $\lambda = L = 2.0$ m. This means the wave number is

$$k = \frac{2\pi}{\lambda} = \frac{2\pi}{2.0\text{ m}} = \pi\text{ rad/m}$$

According to Equation 21.5, the displacement of a medium when two sinusoidal waves superpose to give a standing wave is $D(x,t) = A(x)\cos\omega t$, where $A(x) = 2a\sin kx = A_{\text{max}}\sin kx$. The amplitude function gives the amplitude of oscillation from point to point in the medium. For $x = 10$ cm,

$$A(x = 10 \text{ cm}) = (2.0 \text{ cm})\sin\left[(\pi \text{ rad/m})(0.10 \text{ m})\right] = 0.62 \text{ cm}$$

Similarly, $A(x = 20 \text{ cm}) = 1.18$ cm, $A(x = 30 \text{ cm}) = 1.62$ cm, $A(x = 40 \text{ cm}) = 1.90$ cm, and $A(x = 50 \text{ cm}) = 2.00$ cm.

Assess: Consistent with the above figure, the amplitude of oscillation is a maximum at $x = 0.50$ m.

21.39. Model: A string fixed at both ends forms standing waves.

Solve: **(a)** Three antinodes means the string is vibrating as the $m = 3$ standing wave. The frequency is $f_3 = 3f_1$, so the fundamental frequency is $f_1 = \frac{1}{3}(420 \text{ Hz}) = 140$ Hz. The fifth harmonic will have the frequency $f_5 = 5f_1 = 700$ Hz.

(b) The wavelength of the fundamental mode is $\lambda_1 = 2L = 1.20$ m. The wave speed on the string is $v = \lambda_1 f_1 = (1.20 \text{ m})$ (140 Hz) = 168 m/s. Alternatively, the wavelength of the $n = 3$ mode is $\lambda_3 = \frac{1}{3}(2L) = 0.40$ m, from which $v = \lambda_3 f_3 =$ (0.40 m)(420 Hz) = 168 m/s. The wave speed on the string is given by

$$v = \sqrt{\frac{T_S}{\mu}} \Rightarrow T_S = \mu v^2 = (0.0020 \text{ kg/m})(168 \text{ m/s})^2 = 56 \text{ N}$$

Assess: You must remember to use the linear density in SI units of kg/m. Also, the speed is the same for all modes, but you must use a matching λ and f to calculate the speed.

21.41. Model: The stretched string with both ends fixed forms standing waves.

Visualize:

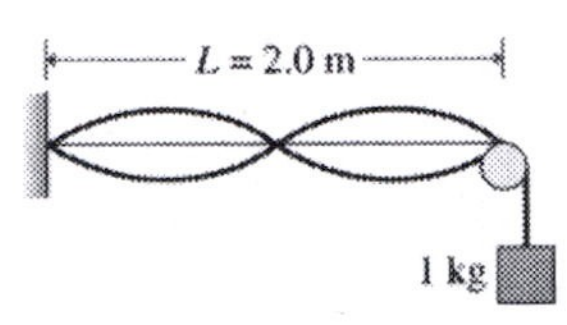

Solve: The astronauts have created a stretched string whose vibrating length is $L = 2.0$ m. The weight of the hanging mass creates a tension $T_S = Mg$ in the string, where $M = 1.0$ kg. As a consequence, the wave speed on the string is

$$v = \sqrt{\frac{T_S}{\mu}} = \sqrt{\frac{Mg}{\mu}}$$

where μ = (0.0050 kg)/(2.5 m) = 0.0020 kg/m is the linear density. The astronauts then observe standing waves at frequencies of 64 Hz and 80 Hz. The first is *not* the fundamental frequency of the string because 80 Hz $\neq$ 2 $\times$ 64 Hz. But we can easily show that both are multiples of 16 Hz: 64 Hz = $4f_1$ and 80 Hz = $5f_1$. Both frequencies are also multiples of 8 Hz. But 8 Hz cannot be the fundamental frequency because, if it were, there would be a standing wave resonance at 9(8 Hz) = 72 Hz. So the fundamental frequency is f_1 = 16 Hz. The fundamental wavelength is $\lambda_1 = 2L = 4.0$ m. Thus, the wave speed on the string is $v = \lambda_1 f_1 = 64.0$ m/s. Now we can find g on Planet X:

$$v = \sqrt{\frac{Mg}{\mu}} \Rightarrow g = \frac{\mu}{M}v^2 = \frac{0.0020 \text{ kg/m}}{1.0 \text{ kg}}(64 \text{ m/s})^2 = 8.2 \text{ m/s}^2$$

21.53. Model: A stretched wire, which is fixed at both ends, forms a standing wave whose fundamental frequency $f_{1\text{ wire}}$ is the same as the fundamental frequency $f_{1\text{ open-closed}}$ of the open-closed tube. The two frequencies are the same because the oscillations in the wire drive oscillations of the air in the tube.

Visualize:

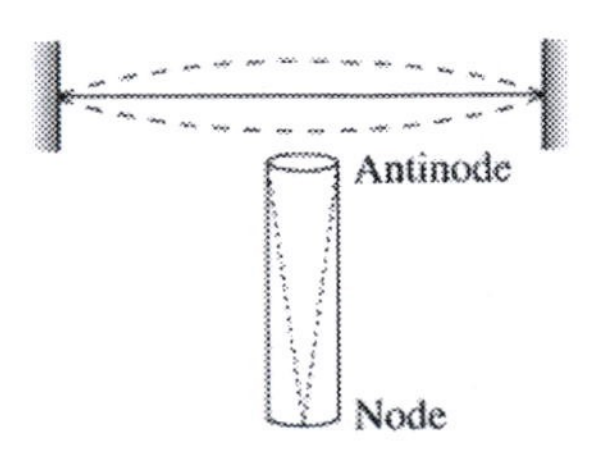

Solve: The fundamental frequency in the wire is

$$f_{1\text{ wire}} = \frac{v_{\text{wire}}}{2L_{\text{wire}}} = \frac{1}{2L_{\text{wire}}}\sqrt{\frac{T_S}{\mu}} = \frac{1}{(1.0\text{ m})}\sqrt{\frac{440\text{ N}}{(0.0010\text{ kg}/0.050\text{ m})}} = 469\text{ Hz}$$

The fundamental frequency in the open-closed tube is

$$f_{1\text{ open-closed}} = 469\text{ Hz} = \frac{v_{\text{air}}}{4L_{\text{tube}}} = \frac{340\text{ m/s}}{4L_{\text{tube}}} \Rightarrow L_{\text{tube}} = \frac{340\text{ m/s}}{4(469\text{ Hz})} = 0.181\text{ m} \approx 18\text{ cm}$$

21.57. Model: Model the tunnel as an open-closed tube.

Visualize: We are given $v = 335$ m/s. We would like to use $f_m = m\dfrac{v}{4L}(m = \text{odd})$ to find L, but we need to know m first. Since m takes on only odd values for the open-closed tube the next resonance after m is $m+2$. We are given $f_m = 4.5$ Hz and $f_{m+2} = 6.3$ Hz.

Solve:

$$\frac{f_{m+2}}{f_m} = \frac{(m+2)\dfrac{v}{4L}}{(m)\dfrac{v}{4L}} = \frac{m+2}{m}$$

$$m\left(\frac{f_{m+2}}{f_m}\right) = m+2$$

$$m\left(\frac{f_{m+2}}{f_m} - 1\right) = 2$$

$$m = \frac{2}{\dfrac{f_{m+2}}{f_m} - 1} = \frac{2}{\dfrac{6.3\text{ Hz}}{4.5\text{ Hz}} - 1} = 5$$

Now that we know m we can finish up.

$$f_m = m\frac{v}{4L} \Rightarrow L = m\frac{v}{4f_m} = (5)\frac{335\text{ m/s}}{4(4.5\text{ Hz})} = 93\text{ m}$$

Assess: 93 m seems like a reasonable length for a tunnel.

21.61. Model: Interference occurs according to the difference between the phases of the two waves.
Visualize:

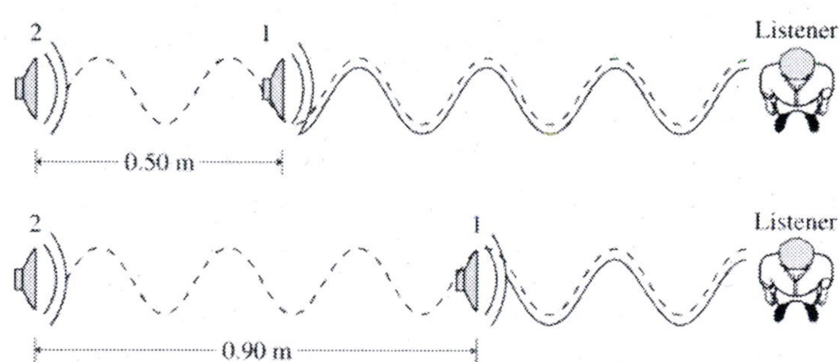

Solve: **(a)** The phase difference between the sound waves from the two speakers is

$$\Delta\phi = 2\pi\frac{\Delta x}{\lambda} + \Delta\phi_0$$

We have a maximum intensity when $\Delta x = 0.50$ m and $\Delta x = 0.90$ m. This means

$$2\pi\frac{(0.50\text{ m})}{\lambda} + \Delta\phi_0 = 2m\pi\text{ rad} \qquad 2\pi\left(\frac{0.90\text{ m}}{\lambda}\right) + \Delta\phi_0 = 2(m+1)\pi\text{ rad}$$

Taking the difference of the above two equations,

$$2\pi\left(\frac{0.40\text{ m}}{\lambda}\right) = 2\pi \Rightarrow \lambda = 0.40\text{ m} \Rightarrow f = \frac{v_{\text{sound}}}{\lambda} = \frac{340\text{ m/s}}{0.40\text{ m}} = 850\text{ Hz}$$

(b) Using again the equations that correspond to constructive interference,

$$2\pi\left(\frac{0.50\text{ m}}{0.40\text{ m}}\right) + \Delta\phi_0 = 2m\pi\text{ rad} \Rightarrow \Delta\phi_0 = \phi_{20} - \phi_{10} = -\frac{\pi}{2}\text{ rad}$$

We have taken $m = 1$ in the last equation. This is because we always specify phase constants in the range $-\pi$ rad to π rad (or 0 rad to 2π rad). $m = 1$ gives $-\frac{1}{2}\pi$ rad (or equivalently, $m = 2$ will give $\frac{3}{2}\pi$ rad).

21.67. Model: The two radio antennas are sources of in-phase, circular waves. The overlap of these waves causes interference.
Visualize:

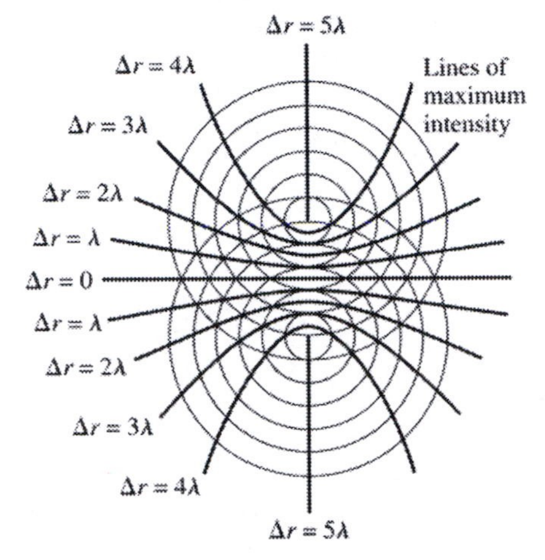

Solve: Maxima occur along lines such that the path difference to the two antennas is $\Delta r = m\lambda$. The 750 MHz = 7.50×10^8 Hz wave has a wavelength $\lambda = c/f = 0.40$ m. Thus, the antenna spacing d = 2.0 m is exactly 5λ. The maximum possible intensity is on the line connecting the antennas, where $\Delta r = d = 5\lambda$. So this is a line of maximum intensity. Similarly, the line that bisects the two antennas is the $\Delta r = 0$ line of maximum intensity. In between, in each of the four quadrants, are four lines of maximum intensity with $\Delta r = \lambda$, 2λ, 3λ, and 4λ. Although we have drawn a fairly accurate picture, you do *not* need to know precisely where these lines are located to know that you *have* to cross them if you walk all the way around the antennas. Thus, you will cross 20 lines where $\Delta r = m\lambda$ and will detect 20 maxima.

21.69. Model: The changing sound intensity is due to the interference of two overlapped sound waves.
Visualize: The listener moving relative to the speakers changes the phase difference between the waves.

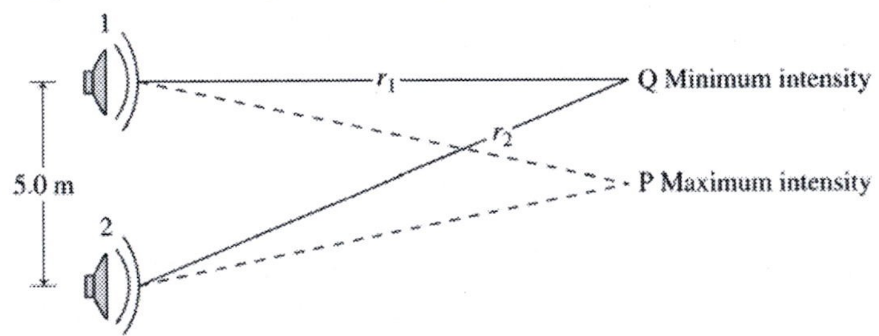

Solve: **(a)** Initially when you are at P, equidistant from the speakers, you hear a sound of maximum intensity. This implies that the two speakers are in phase ($\Delta\phi_0 = 0$). However, on moving to Q you hear a minimum of sound intensity implying that the path length difference from the two speakers to Q is $\lambda/2$. Thus,

$$\tfrac{1}{2}\lambda = \Delta r = \sqrt{(r_1)^2 + (5.0\text{ m})^2} - r_1 = \sqrt{(12.0\text{ m})^2 + (5.0\text{ m})^2} - 12.0\text{ m} = 1.0\text{ m}$$

$$\Rightarrow \lambda = 2.0\text{ m} \Rightarrow f = \frac{v}{\lambda} = \frac{340\text{ m/s}}{2.0\text{ m}} = 170\text{ Hz}$$

(b) At Q, the condition for perfect destructive interference is

$$\Delta\phi = \frac{2\pi(\Delta r)}{\lambda} + 0 \text{ rad} = 2\left(m - \tfrac{1}{2}\right)\pi \text{ rad} \Rightarrow \frac{2\pi\Delta r}{v/f} = 2\left(m - \tfrac{1}{2}\right)\pi \text{ rad}$$

$$\Rightarrow f = \left(m - \tfrac{1}{2}\right)\frac{v}{\Delta r} = \left(m - \tfrac{1}{2}\right)\left(\frac{340 \text{ m/s}}{1.0 \text{ m}}\right)$$

For $m = 1$, 2, and 3, $f_1 = 170$ Hz, $f_2 = 510$ Hz, and $f_3 = 850$ Hz.

21.71. Model: The two radio transmitters are sources of out-of-phase, circular waves. The overlap of these waves causes interference.
Visualize:

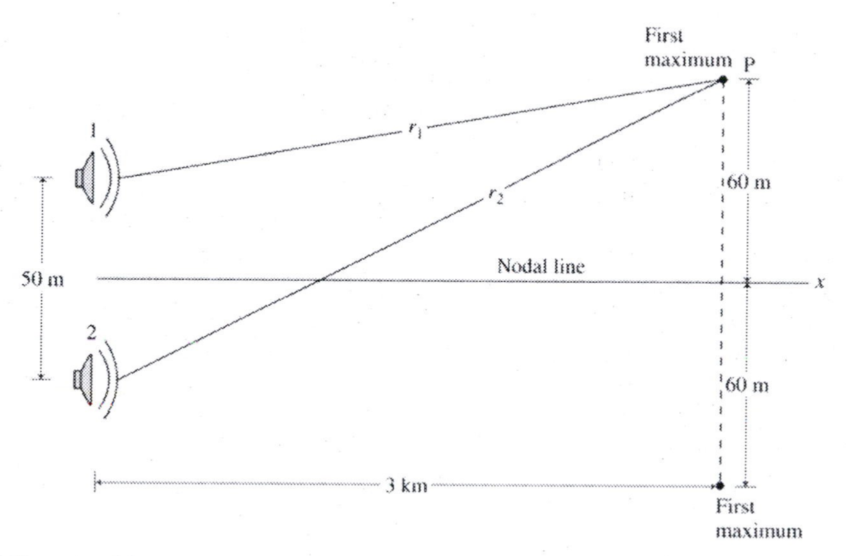

Solve: The phase difference of the waves at point P is given by

$$\Delta\phi = 2\pi\frac{\Delta r}{\lambda} + \Delta\phi_0$$

$$\Delta r = \sqrt{(3000 \text{ m})^2 + (85 \text{ m})^2} - \sqrt{(3000 \text{ m})^2 + (35 \text{ m})^2} = 0.99976 \text{ m}$$

The intensity at P is a maximum. Using $m = 1$ for the first maximum, and $\Delta\phi_0 = \pi$ rad since the transmitters are out of phase, the condition for constructive interference is $\Delta\phi = 2m\pi = 2\pi$. Thus,

$$2\pi \text{ rad} = 2\pi\frac{\Delta r}{\lambda} + \pi \text{ rad} \Rightarrow \lambda = 2\Delta r = 2(0.99976 \text{ m}) \Rightarrow f = \frac{c}{\lambda} = \frac{3\times10^8 \text{ m/s}}{2(0.99976 \text{ m})} = 150 \text{ MHz}$$

21.73. Model: The amplitude is determined by the interference of the two waves.
Solve: **(a)** We have three identical loudspeakers as sources. Δr between speakers 1 and 2 is 1.0 m and $\lambda = 2.0$ m. Thus $\Delta r = \frac{1}{2}\lambda$, which gives perfect destructive interference for in-phase sources. That is, the interference of the waves from loudspeakers 1 and 2 is perfect destructive, leaving only the contribution due to speaker 3. Thus the amplitude is a.
(b) If loudspeaker 2 is moved away by one-half of a wavelength or 1.0 m, then all three waves will reach you in phase. The amplitude of the superposed waves will therefore be maximum and equal to $A = 3a$.
(c) The maximum intensity is $I_{max} = CA^2 = 9Ca^2$. The ratio of the intensity to the intensity of a single speaker is

$$\frac{I_{max}}{I_{single\ speaker}} = \frac{9Ca^2}{Ca^2} = 9$$

21.77. Model: The frequency of the loudspeaker's sound in the back of the pick-up truck is Doppler shifted. As the truck moves away from you, its frequency is decreased.

Solve: Because you hear 8 beats per second as the truck drives away from you, the frequency of the sound from the speaker in the pick-up truck is $f_- = 400\text{ Hz} - 8\text{ Hz} = 392\text{ Hz}$. This frequency is

$$f_- = \frac{f_0}{1 + v_S/v} \Rightarrow 1 + \frac{v_S}{343\text{ m/s}} = \frac{400\text{ Hz}}{392\text{ Hz}} = 1.020408 \Rightarrow v_S = 7.0\text{ m/s}$$

That is, the velocity of the source v_S and hence the pick-up truck is 7.0 m/s.

21.81. Model: The steel wire is under tension and it vibrates with three antinodes.
Solve: When the spring is stretched 8.0 cm, the standing wave on the wire has three antinodes. This means $\lambda_3 = \frac{2}{3}L$ and the tension T_S in the wire is $T_S = k\,(0.080\text{ m})$, where k is the spring constant. For this tension,

$$v_{\text{wire}} = \sqrt{\frac{T_S}{\mu}} \Rightarrow f\lambda_3 = \sqrt{\frac{T_S}{\mu}} \Rightarrow f = \frac{3}{2L}\sqrt{\frac{k(0.08\text{ m})}{\mu}}$$

We will let the stretching of the spring be Δx when the standing wave on the wire displays two antinodes. This means $\lambda_2 = L$ and $T_S' = kx$. For the tension T_S',

$$v'_{\text{wire}} = \sqrt{\frac{T_S'}{\mu}} \Rightarrow f\lambda_2 = \sqrt{\frac{T_S'}{\mu}} \Rightarrow f = \frac{1}{L}\sqrt{\frac{k\Delta x}{\mu}}$$

The frequency f is the same in the above two situations because the wire is driven by the same oscillating magnetic field. Now, equating the two frequency equations,

$$\frac{1}{L}\sqrt{\frac{k\Delta x}{\mu}} = \frac{3}{2L}\sqrt{\frac{k(0.080\text{ m})}{\mu}} \Rightarrow \Delta x = 0.18\text{ m} = 18\text{ cm}$$

22

WAVE OPTICS

Exercises and Problems

22.1. Model: Two closely spaced slits produce a double-slit interference pattern.
Visualize: The interference pattern looks like the photograph of Figure 22.3(b). It is symmetrical with the $m = 2$ fringes on both sides of and equally distant from the central maximum.
Solve: The bright fringes occur at angles θ_m such that

$$d\sin\theta_m = m\lambda \qquad m = 0, 1, 2, 3, \ldots$$

$$\Rightarrow \sin\theta_2 = \frac{2\left(500\times10^{-9}\text{ m}\right)}{\left(50\times10^{-6}\text{ m}\right)} = 0.02 \Rightarrow \theta_2 = 0.020\text{ rad} = 0.020\text{ rad}\times\frac{180°}{\pi\text{ rad}} = 1.15°$$

22.7. Model: Two closely spaced slits produce a double-slit interference pattern.
Visualize: The interference pattern looks like the photograph of Figure 22.3(b).
Solve: The dark fringes are located at positions given by Equation 22.9:

$$y'_m = \left(m+\tfrac{1}{2}\right)\frac{\lambda L}{d} \qquad m = 0, 1, 2, 3, \ldots$$

$$\Rightarrow y'_5 - y'_1 = \left(5+\tfrac{1}{2}\right)\frac{\lambda L}{d} - \left(1+\tfrac{1}{2}\right)\frac{\lambda L}{d} \Rightarrow 6.0\times10^{-3}\text{ m} = \frac{4\lambda(60\times10^{-2}\text{ m})}{0.20\times10^{-3}\text{ m}} \Rightarrow \lambda = 500\text{ nm}$$

22.13. Model: A diffraction grating produces an interference pattern.
Visualize: The interference pattern looks like the diagram of Figure 22.8.
Solve: The bright interference fringes are given by

$$d\sin\theta_m = m\lambda \qquad m = 0, 1, 2, 3, \ldots$$

The slit spacing is $d = 1\text{ mm}/500 = 2.00\times10^{-6}\text{ m}$ and $m = 1$. For the red and blue light,

$$\theta_{1\text{ red}} = \sin^{-1}\left(\frac{656\times10^{-9}\text{ m}}{2.00\times10^{-6}\text{ m}}\right) = 19.15° \qquad \theta_{1\text{ blue}} = \sin^{-1}\left(\frac{486\times10^{-9}\text{ m}}{2.00\times10^{-6}\text{ m}}\right) = 14.06°$$

The distance between the fringes, then, is $\Delta y = y_{1\text{ red}} - y_{1\text{ blue}}$ where

$$y_{1\text{ red}} = (1.5\text{ m})\tan19.15° = 0.521\text{ m}$$
$$y_{1\text{ blue}} = (1.5\text{ m})\tan14.06° = 0.376\text{ m}$$

So, $\Delta y = 0.145\text{ m} = 14.5\text{ cm}$.

22.21. Model: The crack in the cave is like a single slit that causes the ultrasonic sound beam to diffract.
Visualize:

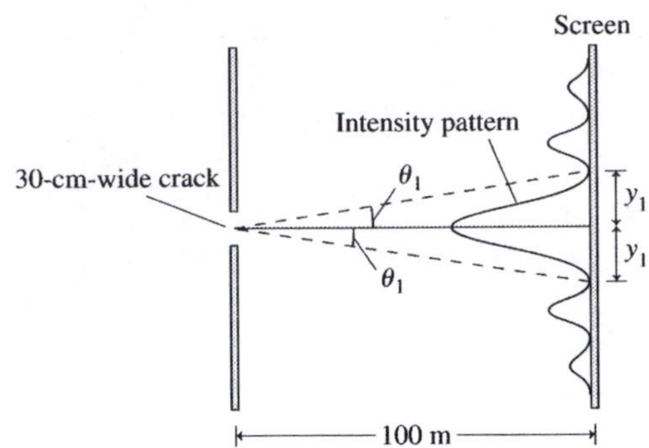

Solve: The wavelength of the ultrasound wave is

$$\lambda = \frac{340 \text{ m/s}}{30 \text{ kHz}} = 0.0113 \text{ m}$$

Using the condition for complete destructive interference with $p = 1$,

$$a\sin\theta_1 = \lambda \Rightarrow \theta_1 = \sin^{-1}\left(\frac{0.0113 \text{ m}}{0.30 \text{ m}}\right) = 2.165°$$

From the geometry of the diagram, the width of the sound beam is

$$w = 2y_1 = 2(100 \text{ m} \times \tan\theta_1) = 200 \text{ m} \times \tan 2.165° = 7.6 \text{ m}$$

Assess: The small-angle approximation is almost always valid for the diffraction of light, but may not be valid for the diffraction of sound waves, which have a much larger wavelength.

22.25. Model: Light passing through a circular aperture leads to a diffraction pattern that has a circular central maximum surrounded by a series of secondary bright fringes.
Visualize: The intensity pattern will look like Figure 22.15.
Solve: From Equation 22.24,

$$L = \frac{Dw}{2.44\lambda} = \frac{(0.12\times10^{-3} \text{ m})(1.0\times10^{-2} \text{ m})}{2.44(633\times10^{-9} \text{ m})} = 78 \text{ cm}$$

22.29. Model: An interferometer produces a new maximum each time L_2 increases by $\frac{1}{2}\lambda$ causing the path-length difference Δr to increase by λ.
Visualize: Please refer to the interferometer in Figure 22.20.
Solve: For sodium light of the longer wavelength (λ_1) and of the shorter wavelength (λ_2),

$$\Delta L = m\frac{\lambda_1}{2} \qquad \Delta L = (m+1)\frac{\lambda_2}{2}$$

We want the same path difference $2(L_2 - L_1)$ to correspond to one extra wavelength for the sodium light of shorter wavelength (λ_2). Thus, we combine the two equations to obtain:

$$m\frac{\lambda_1}{2} = (m+1)\frac{\lambda_2}{2} \Rightarrow m(\lambda_1 - \lambda_2) = \lambda_2 \Rightarrow m = \frac{\lambda_2}{\lambda_1 - \lambda_2} = \frac{589.0 \text{ nm}}{589.6 \text{ nm} - 589.0 \text{ nm}} = 981.67 \cong 982$$

Thus, the distance by which M_2 is to be moved is

$$\Delta L = m\frac{\lambda_1}{2} = 982\left(\frac{589.6 \text{ nm}}{2}\right) = 0.2895 \text{ mm}$$

22.33. Model: Two closely spaced slits produce a double-slit interference pattern.
Solve: The light intensity of a double-slit interference pattern at a position y on the screen is

$$I_{\text{double}} = 4I_1\cos^2\left(\frac{\pi d}{\lambda L}y\right) = 4I_1\cos^2\left(\frac{y}{\Delta y}\pi\right)$$

where $\Delta y = \lambda L/d = 4.0$ mm is the fringe spacing.

Using this value for $\lambda L/d$, we can find the position on the interference pattern where $I_{\text{double}} = I_1$ as follows:

$$4I_1\cos^2\left(\frac{\pi}{4.0\times10^{-3}\ \text{m}}y\right) = I_1 \Rightarrow \left(\frac{\pi}{4.0\times10^{-3}\ \text{m}}\right)y = \cos^{-1}\left(\frac{1}{2}\right) = \frac{\pi}{3}\ \text{rad} \Rightarrow y = \frac{4.0\times10^{-3}\ \text{m}}{3} = 1.3\ \text{mm}$$

22.37. Model: A diffraction grating produces an interference pattern.
Visualize: The interference pattern looks like the diagram in Figure 22.8.
Solve: 500 lines per mm on the diffraction grating gives a spacing between the two lines of $d = 1\ \text{mm}/500 = (1\times10^{-3}\ \text{m})/500 = 2.0\times10^{-6}\ \text{m}$. The wavelength diffracted at angle $\theta_m = 30°$ in order m is

$$\lambda = \frac{d\sin\theta_m}{m} = \frac{(2.0\times10^{-6}\ \text{m})\sin30°}{m} = \frac{1000\ \text{nm}}{m}$$

We're told it is *visible* light that is diffracted at 30°, and the wavelength range for visible light is 400–700 nm. Only $m = 2$ gives a visible light wavelength, so $\lambda = 500$ nm.

22.39. Model: Each wavelength of light is diffracted at a different angle by a diffraction grating.
Solve: Light with a wavelength of 501.5 nm creates a first-order fringe at $y = 21.90$ cm. This light is diffracted at angle

$$\theta_1 = \tan^{-1}\left(\frac{21.90\ \text{cm}}{50.00\ \text{cm}}\right) = 23.65°$$

We can then use the diffraction equation $d\sin\theta_m = m\lambda$, with $m = 1$, to find the slit spacing:

$$d = \frac{\lambda}{\sin\theta_1} = \frac{501.5\ \text{nm}}{\sin(23.65°)} = 1250\ \text{nm}$$

The unknown wavelength creates a first order fringe at $y = 31.60$ cm, or at angle

$$\theta_1 = \tan^{-1}\left(\frac{31.60\ \text{cm}}{50.00\ \text{cm}}\right) = 32.29°$$

With the split spacing now known, we find that the wavelength is

$$\lambda = d\sin\theta_1 = (1250\ \text{nm})\sin(32.29°) = 667.8\ \text{nm}$$

Assess: The distances to the fringes and the first wavelength were given to 4 significant figures. Consequently, we can determine the unknown wavelength to 4 significant figures.

22.41. Model: A diffraction grating produces an interference pattern.
Visualize: The interference pattern looks like the diagram of Figure 22.8.
Solve: **(a)** A grating diffracts light at angles $\sin\theta_m = m\lambda/d$. The distance between adjacent slits is $d = 1\ \text{mm}/600 = 1.667\times10^{-6}\ \text{m} = 1667$ nm. The angle of the $m = 1$ fringe is

$$\theta_1 = \sin^{-1}\left(\frac{500\ \text{nm}}{1667\ \text{nm}}\right) = 17.46°$$

The distance from the central maximum to the $m = 1$ bright fringe on a screen at distance L is

$$y_1 = L\tan\theta_1 = (2\ \text{m})\tan17.46° = 0.629\ \text{m}$$

(Note that the small angle approximation is *not* valid for the maxima of diffraction gratings, which almost always have angles > 10°.) There are two $m = 1$ bright fringes, one on either side of the central maximum. The distance between them is $\Delta y = 2y_1 = 1.258\ \text{m} \approx 1.3\ \text{m}$.

(b) The maximum number of fringes is determined by the maximum value of m for which $\sin\theta_m$ does not exceed 1 because there are no physical angles for which $\sin\theta > 1$. In this case,

$$\sin\theta_m = \frac{m\lambda}{d} = \frac{m(500\ \text{nm})}{1667\ \text{nm}}$$

We can see by inspection that $m = 1$, $m = 2$, and $m = 3$ are acceptable, but $m = 4$ would require a physically impossible $\sin\theta_4 > 1$. Thus, there are three bright fringes on either side of the central maximum plus the central maximum itself for a total of seven bright fringes.

22.45. Model: A diffraction grating produces an interference pattern that is determined by both the slit spacing and the wavelength used.

Solve: An 800 line/mm diffraction grating has a slit spacing $d = (1.0\times10^{-3}\text{ m})/800 = 1.25\times10^{-6}\text{ m}$. Referring to Figure P22.45, the angle of diffraction is given by

$$\tan\theta_1 = \frac{y_1}{L} = \frac{0.436\text{ m}}{1.0\text{ m}} = 0.436 \Rightarrow \theta_1 = 23.557° \Rightarrow \sin\theta_1 = 0.400$$

Using the constructive-interference condition $d\sin\theta_m = m\lambda$,

$$\lambda = \frac{d\sin\theta_1}{1} = (1.25\times10^{-6}\text{ m})(0.400) = 500\text{ nm}$$

We can obtain the same value of λ by using the second-order interference fringe. We first obtain θ_2:

$$\tan\theta_2 = \frac{y_2}{L} = \frac{0.436\text{ m} + 0.897\text{ m}}{1.0\text{ m}} = 1.333 \Rightarrow \theta_2 = 53.12° \Rightarrow \sin\theta_2 = 0.800$$

Using the constructive-interference condition,

$$\lambda = \frac{d\sin\theta_2}{2} = \frac{(1.25\times10^{-6}\text{ m})(0.800)}{2} = 500\text{ nm}$$

Assess: Calculations with the first-order and second-order fringes of the interference pattern give the same value for the wavelength.

22.53. Model: A narrow slit produces a single-slit diffraction pattern.

Visualize: The dark fringes in this diffraction pattern are given by Equation 22.21:

$$y_p = \frac{p\lambda L}{a} \qquad p = 1, 2, 3, \ldots$$

We note that the first minimum in the figure is 0.50 cm away from the central maximum. We are given a = 0.02 nm and $L = 1.5$ m.

Solve: Solve the above equation for λ.

$$\lambda = \frac{y_p a}{pL} = \frac{(0.50\times10^{-2}\text{ m})(0.20\times10^{-3}\text{ m})}{(1)(1.5\text{ m})} = 670\text{ nm}$$

Assess: 670 nm is in the visible range.

22.57. Model: Light passing through a circular aperture leads to a diffraction pattern that has a circular central maximum surrounded by a series of secondary bright fringes.

Solve: **(a)** Because the visible spectrum spans wavelengths from 400 nm to 700 nm, we take the average wavelength of sunlight to be 550 nm.

(b) Within the small-angle approximation, the width of the central maximum is

$$w = 2.44\frac{\lambda L}{D} \Rightarrow (1\times10^{-2}\text{ m}) = (2.44)\frac{(550\times10^{-9}\text{ m})(3\text{ m})}{D} \Rightarrow D = 4.03\times10^{-4}\text{ m} = 0.40\text{ mm}$$

22.59. Model: The laser beam is diffracted through a circular aperture.
Visualize:

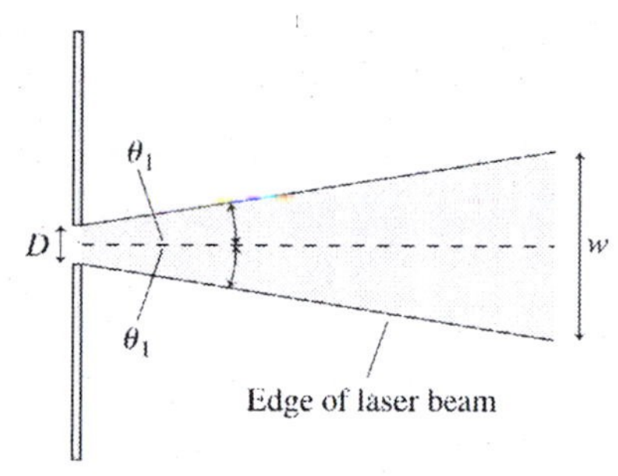

Solve: **(a)** No. The laser light emerges through a circular aperture at the end of the laser. This aperture causes diffraction, hence the laser beam must gradually spread out. The diffraction angle is small enough that the laser beam *appears* to be parallel over short distances. But if you observe the laser beam at a large distance it is easy to see that the diameter of the beam is slowly increasing.
(b) The position of the first minimum in the diffraction pattern is more or less the "edge" of the laser beam. For diffraction through a circular aperture, the first minimum is at an angle

$$\theta_1 = \frac{1.22\lambda}{D} = \frac{1.22\left(633\times10^{-9}\text{ m}\right)}{0.0010\text{ m}} = 7.72\times10^{-4}\text{ rad} = 0.044°$$

(c) The diameter of the laser beam is the width of the diffraction pattern:

$$w = \frac{2.44\lambda L}{D} = \frac{2.44\left(633\times10^{-9}\text{ m}\right)(3\text{ m})}{0.0010\text{ m}} = 0.0046\text{ m} = 0.46\text{ cm}$$

(d) At $L = 1$ km = 1000 m, the diameter is

$$w = \frac{2.44\lambda L}{D} = \frac{2.44\left(633\times10^{-9}\text{ m}\right)(1000\text{ m})}{0.0010\text{ m}} = 1.5\text{ m}$$

22.63. Model: An interferometer produces a new maximum each time L_2 increases by $\frac{1}{2}\lambda$ causing the path-length difference Δr to increase by λ.
Visualize: Please refer to the interferometer in Figure 22.20.
Solve: The path-length difference between the two waves is $\Delta r = 2L_2 - 2L_1$. The condition for constructive interference is $\Delta r = m\lambda$, hence constructive interference occurs when

$$2(L_2 - L_1) = m\lambda \Rightarrow L_2 - L_1 = \tfrac{1}{2}m\lambda_2 = 1200\left(\tfrac{1}{2}\lambda\right) = 600\lambda$$

where λ = 632.8 nm is the wavelength of the helium-neon laser. When the mirror M_2 is moved back and a hydrogen discharge lamp is used, 1200 fringes shift again. Thus,

$$L_2' - L_1 = 1200\left(\tfrac{1}{2}\lambda'\right) = 600\lambda'$$

where $\lambda' = 656.5$ nm. Subtracting the two equations,

$$(L_2 - L_1) - (L_2' - L_1) = 600(\lambda - \lambda') = 600\left(632.8\times10^{-9}\text{ m} - 656.5\times10^{-9}\text{ m}\right)$$

$$\Rightarrow L_2' = L_2 + 14.2\times10^{-6}\text{ m}$$

That is, M_2 is now 14.2 μm closer to the beam splitter.

22.67. Model: The arms of the interferometer are of equal length, so without the crystal the output would be bright.
Visualize: We need to consider how many more wavelengths fit in the electro-optic crystal than would have occupied that space (6.70 μm) without the crystal; if it is an integer then the interferometer will produce a bright output; if it is a half-integer then the interferometer will produce a dark output. But the wavelength we need to consider is the wavelength inside the crystal, not the wavelength in air.

$$\lambda_n = \frac{\lambda}{n}$$

We are told the initial n with no applied voltage is 1.522, and the wavelength in air is $\lambda = 1.000\ \mu$m.

Solve: The number of wavelengths that would have been in that space without the crystal is

$$\frac{6.70\ \mu\text{m}}{1.000\ \mu\text{m}} = 6.70$$

(a) With the crystal in place (and $n = 1.522$) the number of wavelengths in the crystal is

$$\frac{6.70\ \mu\text{m}}{1.000\ \mu\text{m}/1.522} = 10.20$$

$$10.20 - 6.70 = 3.50$$

which shows there are a half-integer number more wavelengths with the crystal in place than if it weren't there. Consequently the output is dark with the crystal in place but no applied voltage.
(b) Since the output was dark in the previous part, we want it to be bright in the new case with the voltage on. That means we want to have just one half more extra wavelengths in the crystal (than if it weren't there) than we did in the previous part. That is, we want 4.00 extra wavelengths in the crystal instead of 3.5, so we want 6.70 + 4.00 = 10.70 wavelengths in the crystal.

$$\frac{6.70\ \mu\text{m}}{1.000\ \mu\text{m}/n} = 10.70 \quad \Rightarrow \quad n = \frac{10.70(1.000\ \mu\text{m})}{6.70\ \mu\text{m}} = 1.597$$

Assess: It seems reasonable to be able to change the index of refraction of a crystal from 1.522 to 1.597 by applying a voltage.

Ray Optics

23

Exercises and Problems

23.1. Model: Light rays travel in straight lines.
Solve: **(a)** The time is

$$t = \frac{\Delta x}{c} = \frac{1.0 \text{ m}}{3\times10^{8} \text{ m/s}} = 3.3\times10^{-9} \text{ s} = 3.3 \text{ ns}$$

(b) The refractive indices for water, glass, and cubic zirconia are 1.33, 1.50, and 1.96, respectively. In a time of 3.33 ns, light will travel the following distance in water:

$$\Delta x_{\text{water}} = v_{\text{water}} t = \left(\frac{c}{n_{\text{water}}}\right) t = \left(\frac{3\times10^{8} \text{ m/s}}{1.33}\right)\left(3.33\times10^{-9} \text{ s}\right) = 0.75 \text{ m}$$

Likewise, the distances traveled in the glass and cubic zirconia are $\Delta x_{\text{glass}} = 0.667$ m and $\Delta x_{\text{cubic zirconia}} = 0.458$ m.

Assess: The higher the refractive index of a medium, the slower the speed of light and hence smaller the distance it travels in that medium in a given time.

23.3. Model: Light rays travel in straight lines. The light source is a point source.
Visualize:

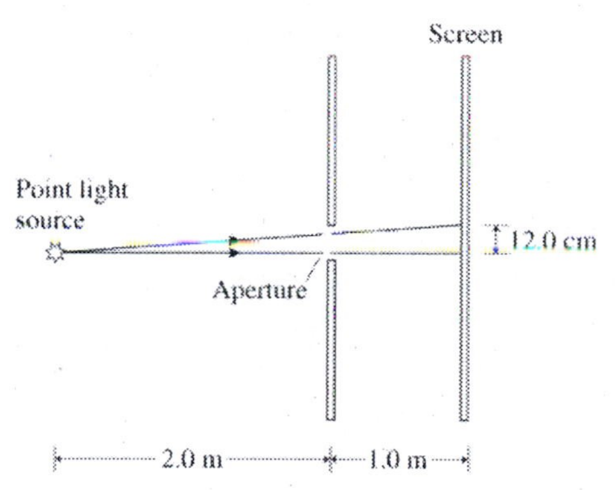

Solve: Let w be the width of the aperture. Then from the geometry of the figure,

$$\frac{w}{2.0 \text{ m}} = \frac{12.0 \text{ cm}}{2.0 \text{ m} + 1.0 \text{ m}} \Rightarrow w = 8.0 \text{ cm}$$

23.9. Model: Light rays travel in straight lines and follow the law of reflection.
Visualize:

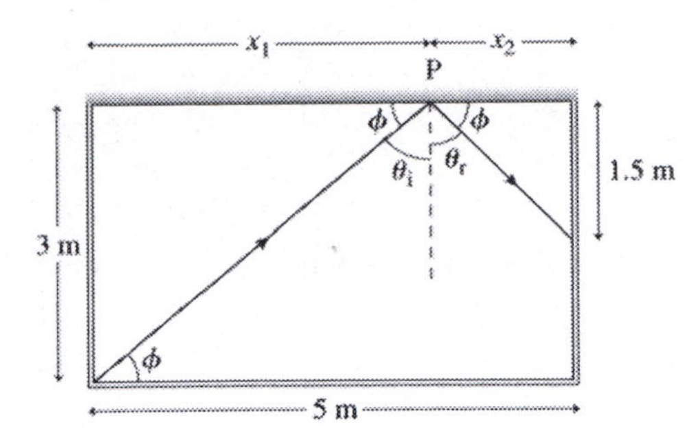

To determine the angle ϕ, we must know the point P on the mirror where the ray is incident. P is a distance x_2 from the far wall and a horizontal distance x_1 from the laser source. The ray from the source must strike P so that the angle of incidence θ_i is equal to the angle of reflection θ_r.
Solve: From the geometry of the diagram,

$$\tan\phi = \frac{1.5\text{ m}}{x_2} = \frac{3\text{ m}}{x_1} \qquad x_1 + x_2 = 5\text{ m}$$

$$\Rightarrow \frac{1.5\text{ m}}{5\text{ m} - x_1} = \frac{3\text{ m}}{x_1} \Rightarrow (1.5\text{ m})x_1 = 15\text{ m}^2 - (3\text{ m})x_1 \Rightarrow x_1 = \frac{10}{3}\text{ m}$$

$$\Rightarrow \tan\phi = \frac{3\text{ m}}{x_1} = \frac{9}{10} = 0.90 \Rightarrow \phi = 42°$$

23.11. Model: Use the ray model of light and the law of reflection.
Visualize:

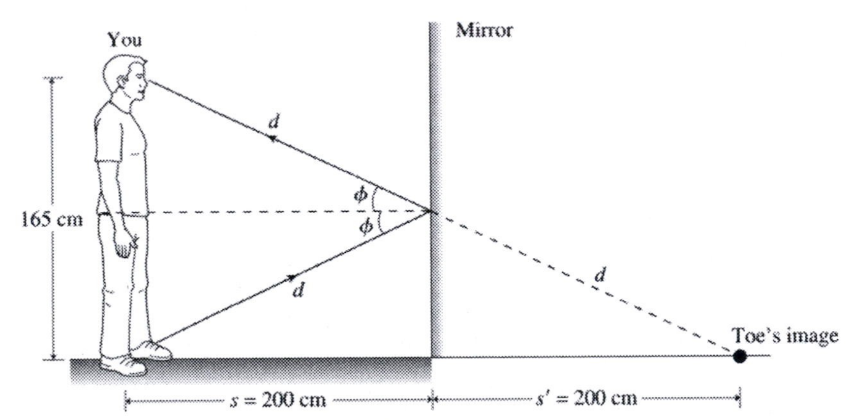

We only need one ray of light that leaves your toes and reflects in your eye.
Solve: From the geometry of the diagram, the distance from your eye to the toes' image is

$$2d = \sqrt{(400\text{ cm})^2 + (165\text{ cm})^2} = 433\text{ cm}$$

Assess: The light appears to come from your toes' image.

23.17. Model: Use the ray model of light. For an angle of incidence greater than the critical angle, the ray of light undergoes total internal reflection.
Visualize:

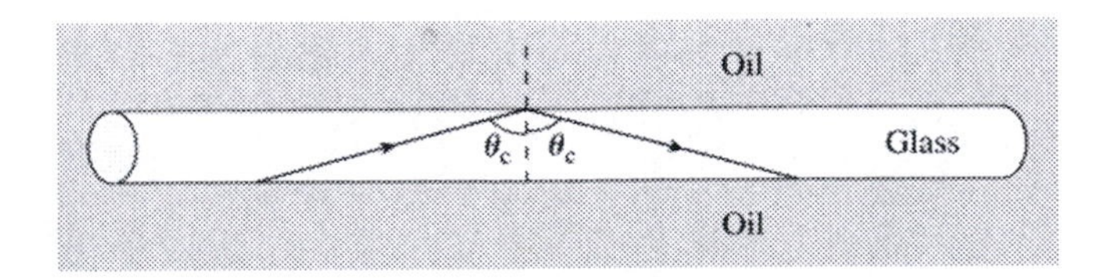

Solve: The critical angle of incidence is given by Equation 23.9:

$$\theta_c = \sin^{-1}\left(\frac{n_{\text{oil}}}{n_{\text{glass}}}\right) = \sin^{-1}\left(\frac{1.46}{1.50}\right) = 76.7°$$

Assess: The critical angle exists because $n_{\text{oil}} < n_{\text{glass}}$.

23.19. Model: Represent the beetle as a point source and use the ray model of light.
Visualize:

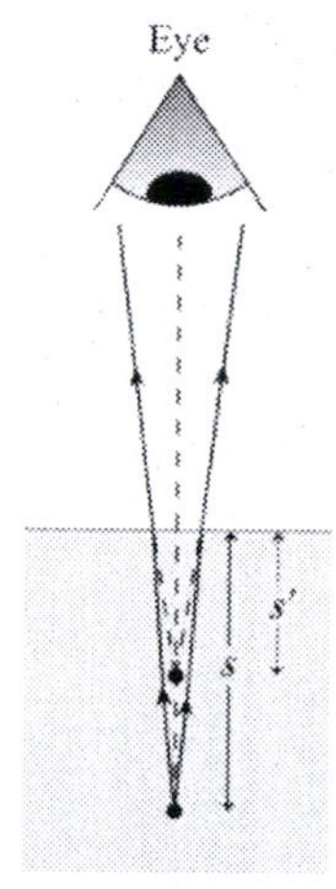

Paraxial rays from the beetle refract into the air and then enter into the observer's eye. The rays in the air when extended into the plastic appear to be coming from the beetle at a shallower location, a distance s' from the plastic-air boundary.
Solve: The actual object distance is s and the image distance is $s' = 2.0$ cm. Using Equation 23.13,

$$s' = \frac{n_2}{n_1}s = \frac{n_{\text{air}}}{n_{\text{plastic}}}s \Rightarrow 2.0\text{ cm} = \frac{1.0}{1.59}s \Rightarrow s = 3.2\text{ cm}$$

Assess: The beetle is much deeper in the plastic than it appears to be.

23.23. Model: Use the ray model of light and the phenomenon of dispersion.
Visualize:

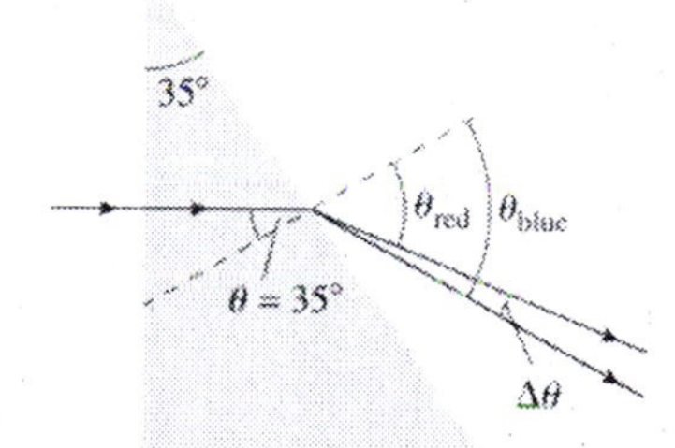

Solve: **(a)** From the graph in Figure 23.29, we estimate the index of refraction for the red light (656 nm) to be $n_{\text{red}} = 1.572$ and for the blue light (456 nm) to be $n_{\text{blue}} = 1.587$.
(b) The angle of incidence onto the rear of the prism is 35°. Using these values for the refractive index and Snell's law,

$$n_{\text{red}}\sin 35° = n_{\text{air}}\sin\theta_{\text{red}} \Rightarrow \theta_{\text{red}} = \sin^{-1}\left(\frac{1.572\sin 35°}{1.0}\right) = 64.4°$$

$$n_{\text{blue}}\sin 35° = n_{\text{air}}\sin\theta_{\text{blue}} \Rightarrow \theta_{\text{blue}} = \sin^{-1}\left(\frac{1.587\sin 35°}{1.0}\right) = 65.5°$$

$$\Rightarrow \Delta\theta = \theta_{\text{blue}} - \theta_{\text{red}} = 1.1°$$

23.27. Model: Use ray tracing to locate the image.
Solve:

The figure shows the ray-tracing diagram using the steps of Tactics Box 23.2. You can see from the diagram that the image is in the plane where the three special rays converge. The image is located at $s' = 15$ cm to the right of the converging lens, and is inverted.

23.31. Model: Assume the planoconvex lens is a thin lens.
Solve: If the object is on the left, then the first surface has $R_1 = \infty$ and the second surface has $R_2 = -40$ cm (concave toward the object). The index of refraction of polystyrene plastic is 1.59, so the lensmaker's equation is

$$\frac{1}{f} = (n-1)\left(\frac{1}{R_1} - \frac{1}{R_2}\right) = (1.59-1)\left(\frac{1}{\infty} - \frac{1}{-40 \text{ cm}}\right) \Rightarrow \frac{1}{f} = \frac{0.59}{40 \text{ cm}} \Rightarrow f = 68 \text{ cm}$$

23.35. Model: Model the bubble as a point source and consider the paraxial rays that refract from the plastic into the air. The edge of the plastic is a spherical refracting surface.
Visualize:

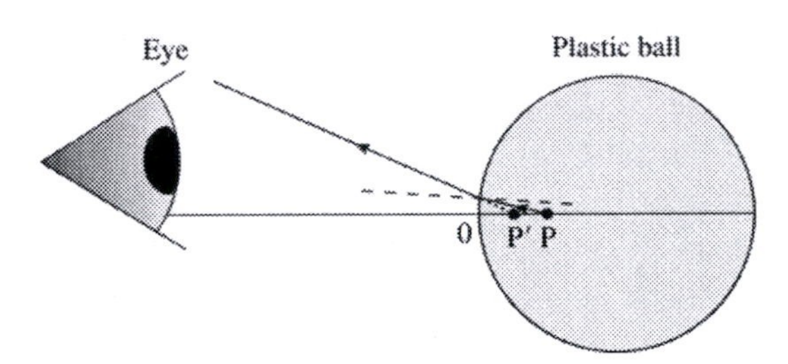

Solve: The bubble is at P, a distance of 2.0 cm from the surface. So, $s = 2.0$ cm. A ray from P after refracting from the plastic-air boundary bends away from the normal axis and enters the eye. This ray appears to come from P′, so the image of P is at P′ and it is a virtual image. Because P faces the concave side of the refracting surface, $R = -4.0$ cm. Furthermore, $n_1 = 1.59$ and $n_2 = 1.0$. Using Equation 23.21,

$$\frac{n_1}{s} + \frac{n_2}{s'} = \frac{n_2 - n_1}{R} \Rightarrow \frac{1.59}{2.0 \text{ cm}} + \frac{1.0}{s'} = \frac{1.0 - 1.59}{-4.0 \text{ cm}} = +\frac{0.59}{4.0 \text{ cm}} = 0.1475 \text{ cm}^{-1}$$

$$\Rightarrow \frac{1}{s'} = 0.1475 \text{ cm}^{-1} - 0.795 \text{ cm}^{-1} \Rightarrow s' = -1.54 \text{ cm}$$

That is, the bubble appears $1.54 \text{ cm} \approx 1.5 \text{ cm}$ beneath the surface.

23.41. Model: Treat the red ball as a point source and use the ray model of light.
Solve: **(a)** Using the law of reflection, we can obtain 3 images of the red ball.
(b) The images of the ball are located at B, C, and D. Relative to the intersection point of the two mirrors, the coordinates of B, C, and D are: B(+1 m, –2 m), C(–1 m, +2 m), and D(+1 m, +2 m).

(c)

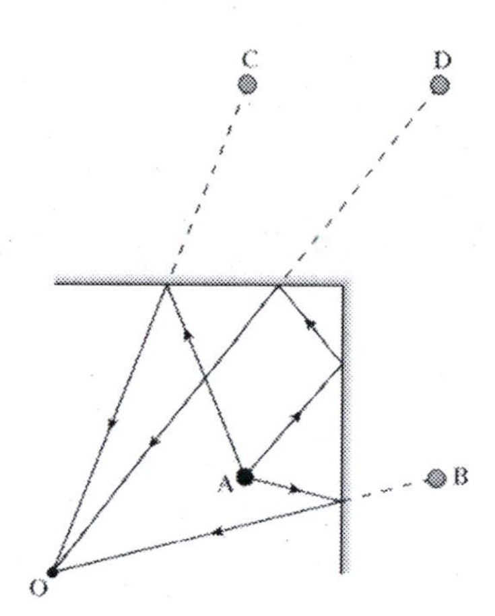

23.49. Model: Use the ray model of light. Light undergoes total internal reflection if it is incident on a boundary at an angle greater than the critical angle.
Visualize:

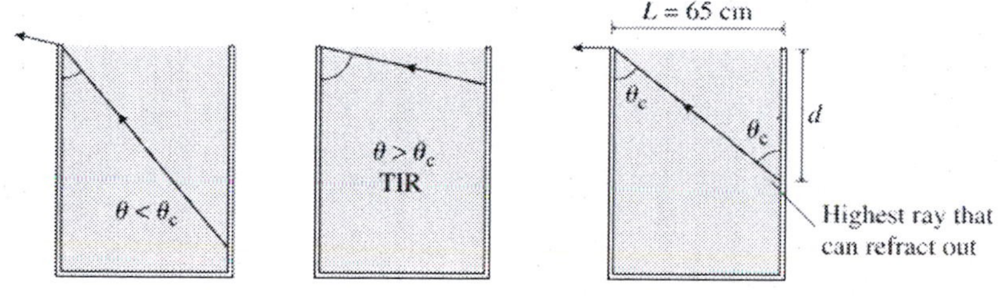

Solve: **(a)** To reach your eye, a light ray must refract through the *top* surface of the water and into the air. You can see in the figure that rays coming from the bottom of the tank are incident on the top surface at fairly small angles, but rays from the marks near the top of the tank are incident at very large angles—greater than the critical angle. These rays undergo total internal reflection in the water and do not exit into the air where they can be seen. Thus you can see the marks from the bottom of the tank upward.
(b) The highest point you can see is the one from which the ray reaches the top surface at the critical angle θ_c. For a water-air boundary, the critical angle is $\theta_c = \sin^{-1}(1/1.33) = 48.75°$. You can see from the figure that the depth of this point is such that

$$\frac{L}{d} = \tan\theta_c \Rightarrow d = \frac{L}{\tan\theta_c} = \frac{65.0\text{ cm}}{\tan(48.75°)} = 57.0\text{ cm}$$

Since the marks are every 10 cm, the high mark you can see is the one at 60 cm.

23.51. Model: Use the ray model of light and the law of refraction. Assume that the laser beam is a ray of light.
Visualize:

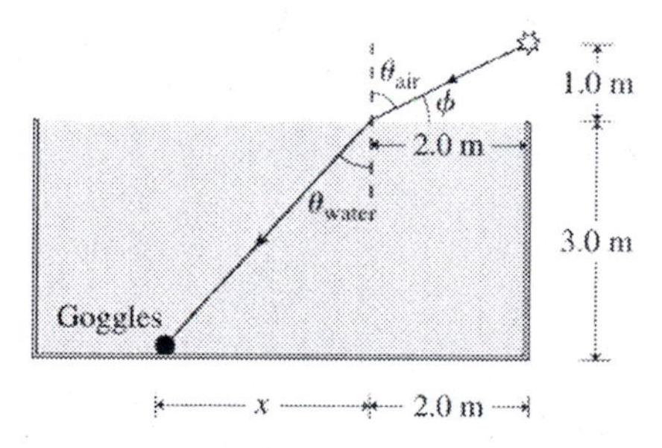

The laser beam enters the water 2.0 m from the edge, undergoes refraction, and illuminates the goggles. The ray of light from the goggles then retraces its path and enters your eyes.
Solve: From the geometry of the diagram,

$$\tan\phi = \frac{1.0\text{ m}}{2.0\text{ m}} \Rightarrow \phi = \tan^{-1}(0.50) = 26.57° \Rightarrow \theta_{air} = 90° - 26.57° = 63.43°$$

Snell's law at the air-water boundary is $n_{\text{air}}\sin\theta_{\text{air}} = n_{\text{water}}\sin\theta_{\text{water}}$. Using the above result,

$$(1.0)\sin 63.43^\circ = 1.33\sin\theta_{\text{water}} \Rightarrow \theta_{\text{water}} = \sin^{-1}\left(\frac{\sin 63.43^\circ}{1.33}\right) = 42.26^\circ$$

Taking advantage of the geometry in the diagram again,

$$\frac{x}{3.0\ \text{m}} = \tan\theta_{\text{water}} \Rightarrow x = (3.0\ \text{m})\tan 42.26^\circ = 2.73\ \text{m}$$

The distance of the goggles from the edge of the pool is $2.73\ \text{m} + 2.0\ \text{m} = 4.73\ \text{m} \approx 4.7\ \text{m}$.

23.53. Model: Use the ray model of light. Assume the bonfire is a point source right at the corner of the lake.
Visualize:

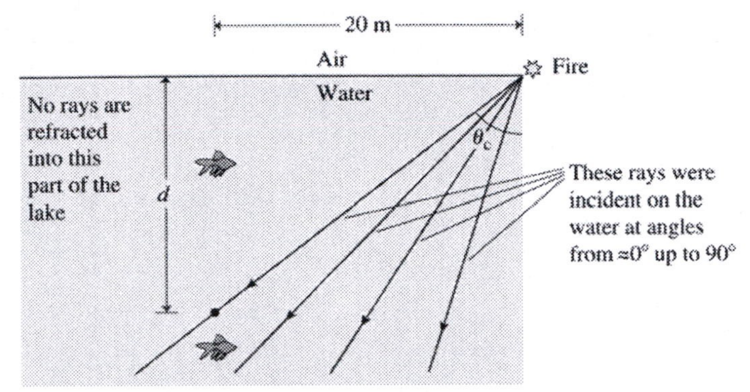

Solve: **(a)** Light rays from the fire enter the lake right at the edge. Even though the rays in air are incident on the surface at a range of angles from ≈0° up to 90°, the larger index of refraction of water causes the rays to travel downward in the water with angles $\le\theta_c$, the critical angle. Some of these rays can reach a fish that is deep in the lake, but a shallow fish out from shore is in the "exclusion zone" that is not reached by any rays from the fire. Thus a fish needs to be deep to see the light from the fire.
(b) The shallowest fish that can see the fire is one that receives light rays refracting into the water at the critical angle θ_c. These are rays that were incident on the water's surface at ≈90°. The critical angle for a water-air boundary is

$$\theta_c = \sin^{-1}\left(\frac{1.00}{1.33}\right) = 48.75^\circ$$

The fish is 20 m from shore, so its depth is

$$d = \frac{20\ \text{m}}{\tan(48.75^\circ)} = 17.5\ \text{m} \approx 18\ \text{m}$$

That is, a fish 20 m from shore must be at least 18 m deep to see the fire.

23.57. Model: Use the ray model of light and the phenomenon of refraction.
Visualize:

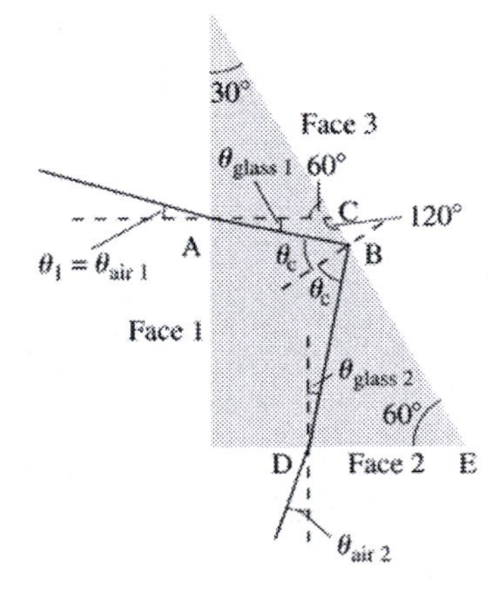

Solve: **(a)** The critical angle θ_c for the glass-air boundary is

$$n_{\text{glass}} \sin\theta_c = n_{\text{air}} \sin 90° \Rightarrow \theta_c = \sin^{-1}\left(\frac{1.0}{1.50}\right) = 41.81°$$

For the triangle ABC,

$$\theta_{\text{glass 1}} + 120° + (90° - \theta_c) = 180° \Rightarrow \theta_{\text{glass 1}} = 180° - 120° - (90° - 41.81°) = 11.81°$$

Having determined $\theta_{\text{glass 1}}$, we can now find $\theta_{\text{air 1}}$ by using Snell's law:

$$n_{\text{air}} \sin\theta_{\text{air 1}} = n_{\text{glass}} \sin\theta_{\text{glass 1}} \Rightarrow \theta_{\text{air 1}} = \sin^{-1}\left(\frac{1.50 \times \sin 11.81°}{1.0}\right) = 17.88°$$

Thus, the smallest angle θ_1 for which a laser beam will undergo TIR on the hypotenuse of this glass prism is 17.9°.
(b) After reflecting from the hypotenuse (face 3) the ray of light strikes the base (face 2) and refracts into the air. From the triangle BDE,

$$(90° - \theta_{\text{glass 2}}) + 60° + (90° - \theta_c) = 180° \Rightarrow \theta_{\text{glass 2}} = 90° + 60° + 90° - 41.81° - 180° = 18.19°$$

Snell's law at the glass-air boundary of face 2 is

$$n_{\text{glass}} \sin\theta_{\text{glass 2}} = n_{\text{air}} \sin\theta_{\text{air 2}} \Rightarrow \theta_{\text{air 2}} = \sin^{-1}\left(\frac{n_{\text{glass}} \sin\theta_{\text{glass 2}}}{n_{\text{air}}}\right) = \sin^{-1}\left(\frac{1.50 \sin 18.19°}{1.0}\right) = 27.9°$$

Thus the ray exits 27.9° left of the normal.

23.61. Model: Assume that the converging lens is a thin lens. Use ray tracing to locate the image.
Solve: **(a)**

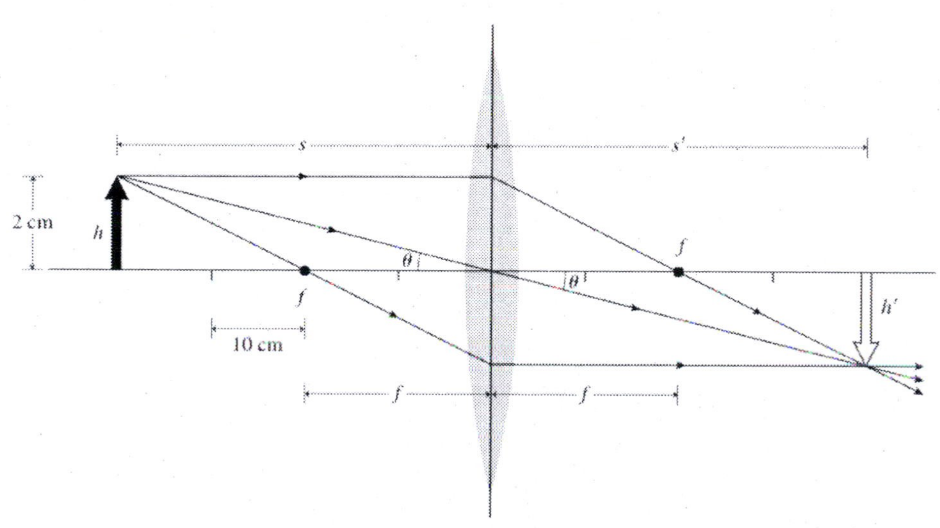

The figure shows the ray-tracing diagram using the steps of Tactics Box 23.2. The three rays after refraction converge to give an image at $s' = 40$ cm. The height of the image is $h' = 2$ cm.
(b) Using the thin-lens formula,

$$\frac{1}{s} + \frac{1}{s'} = \frac{1}{f} \Rightarrow \frac{1}{40\text{ cm}} + \frac{1}{s'} = \frac{1}{20\text{ cm}} \Rightarrow \frac{1}{s'} = \frac{1}{40\text{ cm}} \Rightarrow s' = 40\text{ cm}$$

The image height is obtained from

$$M = -\frac{s'}{s} = -\frac{40\text{ cm}}{40\text{ cm}} = -1$$

The image is inverted and as tall as the object, that is, $h' = 2.0$ cm. The values for h' and s' obtained in parts (a) and (b) agree.

23.65. Model: Use ray tracing to locate the image. Assume the diverging lens is a thin lens.

Solve: (a)

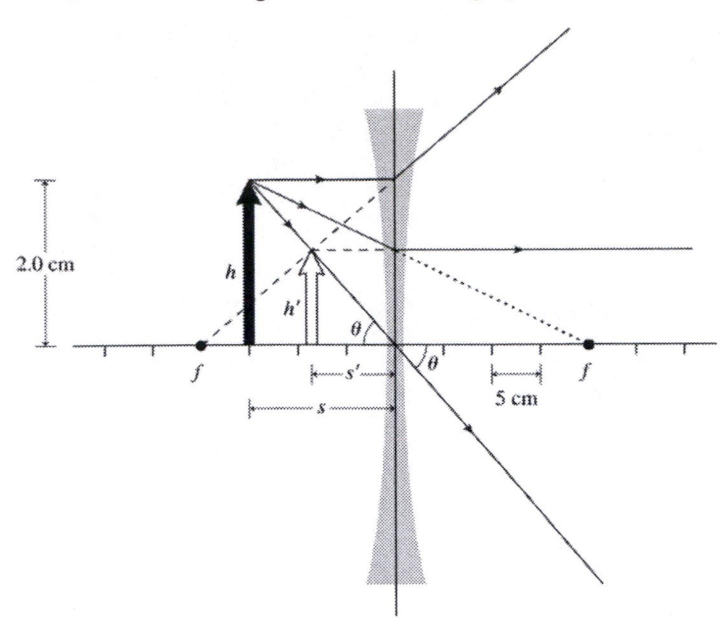

The figure shows the ray-tracing diagram using the steps of Tactics Box 23.3. After refraction, the three special rays do not converge. The rays, on the other hand, appear to meet at a point that is 8.5 cm on the same side of the lens as the object. So $s' = -8.5$ cm. The image is upright and has a height of 1.1 cm.

(b) Using the thin-lens formula,

$$\frac{1}{s}+\frac{1}{s'}=\frac{1}{f} \Rightarrow \frac{1}{15\text{ cm}}+\frac{1}{s'}=\frac{1}{-20\text{ cm}} \Rightarrow \frac{1}{s'}=-\frac{7}{60\text{ cm}} \Rightarrow s'=-\frac{60}{7}\text{ cm}=-8.6\text{ cm}$$

The image height is obtained from

$$M=-\frac{s'}{s}=-\frac{(-60/7\text{ cm})}{15\text{ cm}}=+\frac{4}{7}=0.57$$

Thus, the image is 0.57 times larger than the object, or $h' = Mh = (0.57)(2.0\text{ cm}) = 1.14$ cm. The image is upright because M is positive. These values agree, within measurement accuracy, with those obtained in part (a).

23.73. Visualize: The lens must be a converging lens for this scenario to happen, so we expect f to be positive. In the first case the upright image is virtual $(s_2' < 0)$ and the object must be closer to the lens than the focal point. The lens is then moved backward past the focal point and the image becomes real $(s_2' > 0)$.

$$\frac{1}{s}+\frac{1}{s'}=\frac{1}{f} \Rightarrow f=\frac{ss'}{s+s'}$$

We are given $s_1 = 10\text{cm}$ and $m_1 = 2$.

Solve: Since the first image is virtual, $s' < 0$. We are told the first magnification is $m_1 = 2 = -s_1'/s_1 \Rightarrow s_1' = -20$ cm. We can now find the focal length of the lens.

$$f=\frac{s_1 s_1'}{s_1+s_1'}=\frac{(10\text{ cm})(-20\text{ cm})}{10\text{ cm}-20\text{ cm}}=20\text{ cm}$$

After the lens is moved, $m_2 = -2 = -s_2'/s_2$. Start with the thin lens equation again.

$$\frac{1}{s_2}+\frac{1}{s_2'}=\frac{1}{f}$$

Replace s_2' with $-m_2 s_2$.

$$\frac{1}{s_2}+\frac{1}{-m_2 s_2}=\frac{1}{f}$$

Now solve for s_2.

$$\frac{-m_2 s_2 + s_2}{s_2(-m_2 s_2)} = \frac{1}{f}$$

$$\frac{-m_2 s_2 + s_2}{-m_2 s_2^2} = \frac{1}{f}$$

Cancel one s_2.

$$\frac{m_2 - 1}{m_2 s_2} = \frac{1}{f}$$

Multiply both sides by fs_2.

$$s_2 = f\left(\frac{m_2 - 1}{m_2}\right) = (20\text{ cm})\left(\frac{-2-1}{-2}\right) = 30\text{ cm}$$

The distance the lens moved is $s_2 - s_1 = 30\text{ cm} - 10\text{ cm} = 20\text{ cm}$.

Assess: We knew s_2 needed to be bigger than *f*; it is, and is in a reasonable range. The final answer for the distance the lens moved also seems reasonable.

OPTICAL INSTRUMENTS

24

Exercises and Problems

24.1. Model: Each lens is a thin lens. The image of the first lens is the object for the second lens.

Visualize:

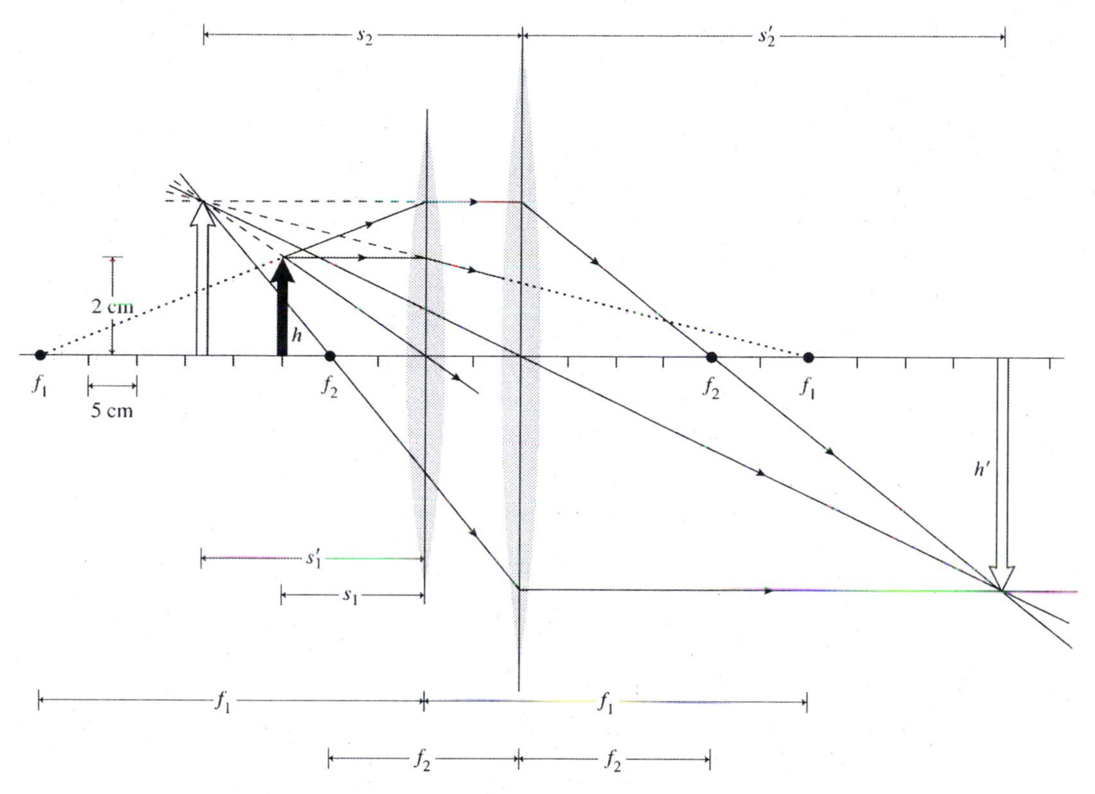

The figure shows the two lenses and a ray-tracing diagram. The ray-tracing shows that the lens combination will produce a real, inverted image behind the second lens.

Solve: (a) From the ray-tracing diagram, we find that the image is ≈ 50 cm from the second lens and the height of the final image is 4.5 cm.

(b) s_1 = 15 cm is the object distance of the first lens. Its image, which is a virtual image, is found from the thin-lens equation:

$$\frac{1}{s_1'} = \frac{1}{f_1} - \frac{1}{s_1} = \frac{1}{40 \text{ cm}} - \frac{1}{15 \text{ cm}} = -\frac{5}{120 \text{ cm}} \Rightarrow s_1' = -24 \text{ cm}$$

The magnification of the first lens is

$$m_1 = -\frac{s_1'}{s_1} = -\frac{(-24 \text{ cm})}{15 \text{ cm}} = 1.6$$

The image of the first lens is now the object for the second lens. The object distance is s_2 = 24 cm + 10 cm = 34 cm. A second application of the thin-lens equation yields:

$$\frac{1}{s_2'} = \frac{1}{f_2} - \frac{1}{s_2} = \frac{1}{20 \text{ cm}} - \frac{1}{34 \text{ cm}} \Rightarrow s_2' = \frac{680 \text{ cm}}{14} = 48.6 \text{ cm}$$

The magnification of the second lens is

$$m_2 = -\frac{s_2'}{s_2} = -\frac{48.6 \text{ cm}}{34 \text{ cm}} = -1.429$$

The combined magnification is $m = m_1 m_2 = (1.6)(-1.429) = -2.286$. The height of the final image is (2.286)(2.0 cm) = 4.57 cm. These calculated values are in agreement with those found in part (a).

24.9. Visualize: First we compute the f-number of the first lens and then the diameter of the second.

Solve:

$$f\text{-number} = \frac{f}{D} = \frac{12 \text{ mm}}{4.0 \text{ mm}} = 3.0$$

Now for the new lens.

$$D = \frac{f}{f\text{-number}} = \frac{18 \text{ mm}}{3.0} = 6.0 \text{ mm}$$

Assess: Given the same f-number, the longer focal length lens has a larger diameter.

24.13. Model: Ignore the small space between the lens and the eye.
Visualize: Refer to Example 24.5, but we want to solve for s', the far point.
Solve:
(a) The power of the lens is negative which means the focal length is negative, so Ellen wears diverging lenses. This is the remedy for myopia.
(b) We want to know where the image should be for an object $s = \infty$ m given $1/f = -1.0 \text{ m}^{-1}$.

$$f = \frac{1}{P} = -1.0 \text{ m}$$

$$\frac{1}{s} + \frac{1}{s'} = \frac{1}{f}$$

When $s = \infty$ m,

$$\frac{1}{\infty \text{ m}} + \frac{1}{s'} = \frac{1}{f} \Rightarrow s' = f = -1.0 \text{ m}$$

So the far point is 100 cm.
Assess: The negative sign on s' is expected because we need the image to be virtual.

24.21. Model: Assume the eyepiece is a simple magnifier with $M_{\text{eye}} = 25 \text{ cm}/f_{\text{eye}}$.

Visualize: $f_{\text{eye}} = 25 \text{ cm}/10 = 2.5$ cm.

Solve:
(a) The magnification of a telescope is

$$M = -\frac{f_{\text{obj}}}{f_{\text{eye}}} = \frac{100 \text{ cm}}{2.5 \text{ cm}} = 40$$

(b)

$$f\text{-number} = \frac{f}{D} = \frac{1.00 \text{ m}}{0.20 \text{ m}} = 5.0$$

Assess: These results are in reasonable ranges for magnification and f-number.

24.23. Model: Two objects are marginally resolvable if the angular separation between the objects, as seen from the lens, is $\alpha = 1.22\lambda / D$.

Solve: Let Δy be the separation between the two light bulbs, and let L be their distance from a telescope. Thus,

$$\alpha = \frac{\Delta y}{L} = \frac{1.22\lambda}{D} \Rightarrow L = \frac{\Delta y\, D}{1.22\lambda} = \frac{(1.0\text{ m})(4.0\times10^{-2}\text{ m})}{1.22(600\times10^{-9}\text{ m})} = 55\text{ km}$$

24.29. Solve: (a) The image location from the first lens is

$$s_1' = \frac{f_1 s_1}{s_1 - f_1} = \frac{(-2.5\text{ cm})(2.5\text{ cm})}{2.5\text{ cm} - (-2.5\text{ cm})} = -1.25\text{ cm}$$

So the image from the first lens is 1.25 cm to the left of the first lens, upright and virtual.
Now, $s_2 = d + 1.25$ cm.
We are told the final image is at infinity: $s_2' = \infty \Rightarrow s_2 = f_2 \Rightarrow f_2 = d + 1.25$ cm

$$d = f_2 - 1.25\text{cm} = 3.75\text{ cm}$$

(b)

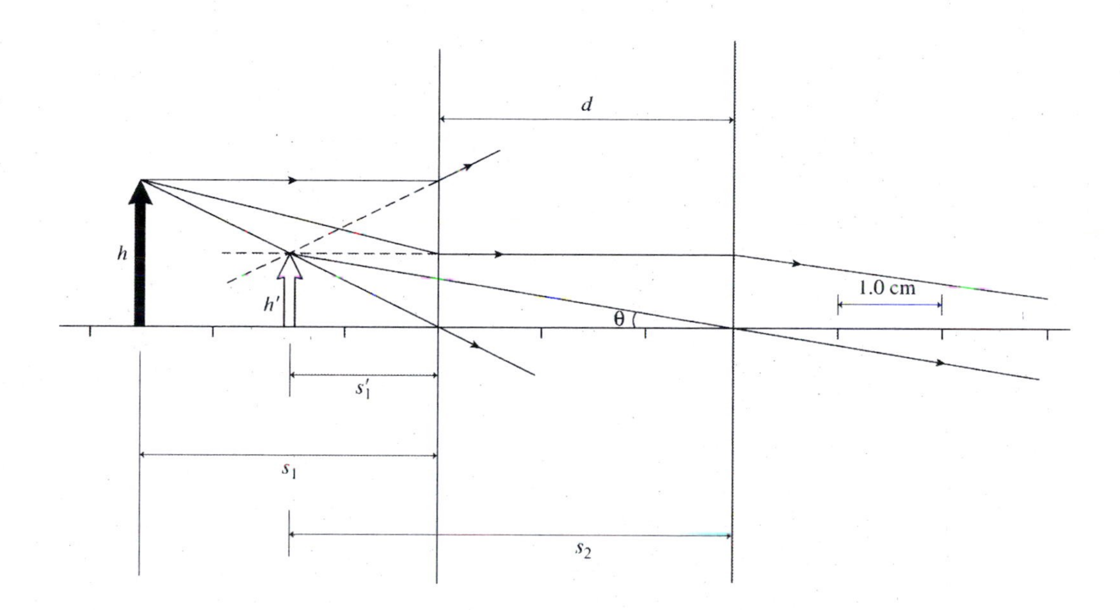

(c)

$$h' = -h\frac{s_1'}{s_1} = 0.50\text{ cm}$$

The angular size is

$$\theta = \tan\frac{h'}{f_2} \approx \frac{h'}{f_2} = \frac{0.50\text{ cm}}{5.0\text{ cm}} = 0.10\text{ rad}$$

(d) If the object were held at the eye's near point, it would subtend:

$$\theta_{\text{NP}} = \frac{h}{25\text{ cm}} = \frac{1.0\text{ cm}}{25\text{ cm}} = 0.040\text{ rad}$$

The angular magnification is

$$M = \frac{\theta}{\theta_{\text{NP}}} = \frac{0.10\text{ rad}}{0.040\text{ rad}} = 2.5$$

Assess: The numerical answers seem to agree with the drawing.

24.31. Visualize: Hard thought shows that if the left focal points for both lenses coincide then the parallel rays before and after the beam splitter are reproduced. The first lens diverges the rays as if they had come from the focal point of the converging lens.

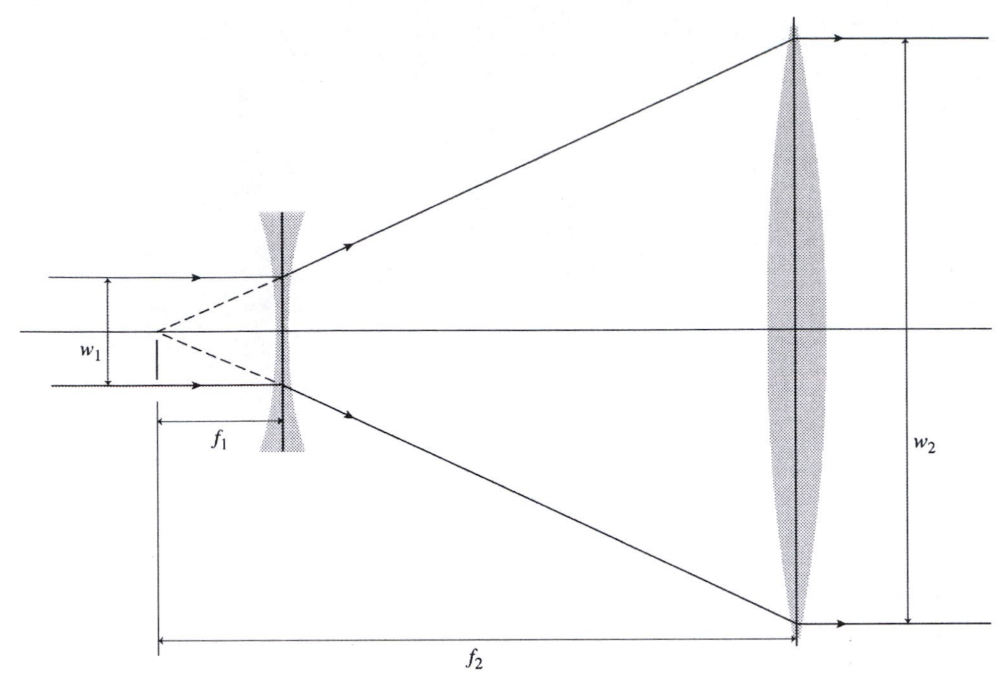

Solve: (a)

$$d = f_2 - |f_1|$$

But since we are given $f_1 < 0$, this is equivalent to

$$d = f_2 + f_1$$

(b) Looking at the similar triangles in the diagram shows that

$$\frac{w_1}{|f_1|} = \frac{w_2}{f_2}$$

$$w_2 = \frac{f_2}{|f_1|} w_1$$

Assess: Figure P24.31 says $f_2 > |f_1|$ and our answer then shows that $w_2 > w_1$ which is the goal of a beam expander.

24.37. Visualize: Use Equation 23.27, the lens makers' equation:

$$\frac{1}{f} = (n-1)\left(\frac{1}{R_1} - \frac{1}{R_2}\right)$$

For a symmetric lens $R_1 = R_2$ and

$$f = \frac{R}{2(n-1)} \quad \text{and} \quad R = 2(n-1)f$$

Also needed will be the magnification of a telescope: $M = -f_{obj}/f_{eye} \Rightarrow f_{eye} = -f_{obj}/M$ (but we will drop the negative sign).

We are given $R_{obj} = 100$ cm and $M = 20$.

Solve:

$$R_{\text{eye}} = 2(n-1)f_{\text{eye}} = 2(n-1)\frac{f_{\text{obj}}}{M} = 2(n-1)\frac{\frac{R_{\text{obj}}}{2(n-1)}}{M} = \frac{R_{\text{obj}}}{M} = \frac{100 \text{ cm}}{20} = 5.0 \text{ cm}$$

Assess: We expect a short focal length and small radius of curvature for telescope eyepieces.

24.43. Model: The width of the central maximum that accounts for a significant amount of diffracted light intensity is inversely proportional to the size of the aperture. The lens is an aperture that focuses light.
Solve: To focus a laser beam, which consists of parallel rays from $s = \infty$, the focal length needs to match the distance to the target: $f = L = 5.0$ cm. The minimum spot size to which a lens can focus is

$$w = \frac{2.44\lambda f}{D} \Rightarrow 5.0\times10^{-6} \text{ m} = \frac{2.44\left(1.06\times10^{-6} \text{ m}\right)\left(5.0\times10^{-2} \text{ m}\right)}{D} \Rightarrow D = 2.6 \text{ cm}.$$

24.49. Model: For a diffraction-limited lens, the minimum focal length is the same size as its diameter. The smallest spot diameter over which you can focus light is $w_{\text{min}} \approx 2.5\lambda$.
Solve: **(a)** The smallest spot size is $w_{\text{min}} \approx 2.5\lambda = 2.5\left(800\times10^{-9} \text{ m}\right) = 2\ \mu\text{m}$.

(b) The total usable area of the optical disk is

$$\pi\left[\left(5.5\times10^{-2} \text{ m}\right)^2 - \left(2\times10^{-2} \text{ m}\right)^2\right] = 0.00825 \text{ m}^2$$

The area of each pit is the area of one bit of information and is $\left[(1.25)(2\ \mu\text{m})\right]^2 = \left[2.5\ \mu\text{m}\right]^2$. The area of 1 byte is 8 times this quantity and the area of 1 megabyte (MB) of information is 10^6 times more. This means the number of megabytes (MB) of data that can be stored on the disk is

$$\frac{0.00825 \text{ m}^2}{8\times\left(2.5\times10^{-6} \text{ m}\right)^2\times10^6 \text{ MB}^{-1}} = 165 \text{ MB}$$

Assess: A memory storage capacity of 165 MB is reasonable.

25

MODERN OPTICS AND MATTER WAVES

Exercises and Problems

25.1. Model: Balmer's formula predicts a series of spectral lines in the hydrogen spectrum.
Solve: Substituting into the formula for the Balmer series,

$$\lambda = \frac{91.18 \text{ nm}}{\left(\frac{1}{2^2} - \frac{1}{n^2}\right)} \Rightarrow \lambda = \frac{91.18 \text{ nm}}{\frac{1}{2^2} - \frac{1}{6^2}} = 410.3 \text{ nm}$$

where $n = 3, 4, 5, 6, \ldots$ and where we have used $n = 6$. Likewise for $n = 8$ and $n = 10$, $\lambda = 389.0$ nm and $\lambda = 379.9$ nm.

25.5. Model: The angles of incidence for which diffraction from parallel planes occurs satisfy the Bragg condition.
Solve: The Bragg condition is $2d\cos\theta_m = m\lambda$. For $m = 1$ and for two different wavelengths,

$$2d\cos\theta_1 = (1)\lambda_1 \qquad 2d\cos\theta_1' = (1)\lambda_1'$$

Dividing these two equations,

$$\frac{\cos\theta_1'}{\cos\theta_1} = \frac{\lambda_1'}{\lambda_1} \Rightarrow \frac{\cos\theta_1'}{\cos 54°} = \frac{0.15 \text{ nm}}{0.20 \text{ nm}} \Rightarrow \theta_1' = \cos^{-1}(0.4408) = 64°$$

25.9. Model: Use the photon model of light.
Solve: The energy of the single photon is

$$E_{\text{photon}} = hf = h\left(\frac{c}{\lambda}\right) = \frac{(6.63\times10^{-34} \text{ Js})(3.0\times10^{8} \text{ m/s})}{1.0\times10^{-6} \text{ m}} = 1.99\times10^{-19} \text{ J}$$

$$\Rightarrow E_{\text{mol}} = N_A E_{\text{photon}} = (6.023\times10^{23})(1.99\times10^{-19} \text{ J}) = 1.2\times10^{5} \text{ J}$$

Assess: Although the energy of a single photon is very small, a mole of photons has a significant amount of energy.

25.13. Solve: **(a)** The baseball's momentum is $p = mv = (0.200 \text{ kg})(30 \text{ m/s}) = 6.0$ kg m/s. The baseball's de Broglie wavelength is

$$\lambda = \frac{h}{p} = \frac{6.63\times10^{-34} \text{ Js}}{6.0 \text{ kg m/s}} = 1.1\times10^{-34} \text{ m}$$

(b) Using $\lambda = h/p = h/mv$, we have

$$v = \frac{h}{m\lambda} = \frac{6.63\times10^{-34} \text{ Js}}{(0.200 \text{ kg})(0.20\times10^{-9} \text{ m})} = 1.7\times10^{-23} \text{ m/s}$$

25.17. Model: A confined particle can have only discrete values of energy.
Solve: From Equation 24.14, the energy of a confined electron is

$$E_n = \frac{h^2}{8mL^2}n^2 \qquad n = 1, 2, 3, 4, \ldots$$

The minimum energy is

$$E_1 = \frac{h^2}{8mL^2} \Rightarrow L = \frac{h}{\sqrt{8mE_1}} = \frac{6.63\times10^{-34}\ \text{Js}}{\sqrt{8(9.11\times10^{-31}\ \text{kg})(1.5\times10^{-18}\ \text{J})}} = 2.0\times10^{-10}\ \text{m} = 0.20\ \text{nm}$$

25.19. Model: The generalized formula of Balmer predicts a series of spectral lines in the hydrogen spectrum.
Solve: (a) The generalized formula of Balmer

$$\lambda = \frac{91.18\ \text{m}}{\left(\frac{1}{m^2} - \frac{1}{n^2}\right)}$$

with $m = 1$ and $n > 1$ accounts for a series of spectral lines. This series is called the Lyman series and the first two members are

$$\lambda_1 = \frac{91.18\ \text{m}}{\left(1 - \frac{1}{2^2}\right)} = 121.6\ \text{nm} \qquad \lambda_2 = \frac{91.18\ \text{nm}}{\left(1 - \frac{1}{3^2}\right)} = 102.6\ \text{nm}$$

For $n = 4$ and $n = 5$, $\lambda_3 = 97.3$ nm and $\lambda_4 = 95.0$ nm .

(b) The Lyman series converges when $n \to \infty$. This means $1/n^2 \to 0$ and $\lambda \to 91.18$ nm.

(c) For a diffraction grating, the condition for bright (constructive interference) fringes is $d\sin\theta_p = p\lambda$, where $p = 1, 2, 3, \ldots$ For first-order diffraction, this equation simplifies to $d\sin\theta = \lambda$. For the first and second members of the Lyman series, the above condition is $d\sin\theta_1 = \lambda_1 = 121.6$ nm and $d\sin\theta_2 = \lambda_2 = 102.6$ nm. Dividing these two equations yields

$$\sin\theta_2 = \left(\frac{102.6\ \text{nm}}{121.6\ \text{nm}}\right)\sin\theta_1 = (0.84375)\sin\theta_1$$

The distance from the center to the first maximum is $y = L\tan\theta$. Thus,

$$\tan\theta_1 = \frac{y_1}{L} = \frac{0.376\ \text{m}}{1.5\ \text{m}} \Rightarrow \theta_1 = 14.072° \Rightarrow \sin\theta_2 = (0.84375)\sin(14.072°) \Rightarrow \theta_2 = 11.84°$$

Applying the position formula once again,

$$y_2 = L\tan\theta_2 = (1.5\ \text{m})\tan(11.84°) = 0.314\ \text{m} = 31.4\ \text{cm}$$

25.21. Model: Use the photon model of light.
Solve: (a) The wavelength is calculated as follows:

$$E_{\text{gamma}} = hf = h\left(\frac{c}{\lambda}\right) \Rightarrow \lambda = \frac{(6.63\times10^{-34}\ \text{Js})(3.0\times10^8\ \text{m/s})}{1.0\times10^{-13}\ \text{J}} = 2.0\times10^{-12}\ \text{m}$$

(b) The energy of a visible-light photon of wavelength 500 nm is

$$E_{\text{visible}} = h\left(\frac{c}{\lambda}\right) = \frac{(6.63\times10^{-34}\ \text{Js})(3.0\times10^8\ \text{m/s})}{500\times10^{-9}\ \text{m}} = 3.978\times10^{-19}\ \text{J}$$

The number of photons n such that $E_{\text{gamma}} = nE_{\text{visible}}$ is

$$n = \frac{E_{\text{gamma}}}{E_{\text{visible}}} = \frac{1.0\times10^{-13}\ \text{J}}{3.978\times10^{-19}\ \text{J}} = 2.5\times10^5$$

25.25. Model: Use the photon model of light and the Bragg condition for diffraction.
Solve: The Bragg condition for the reflection of x-rays from a crystal is $2d\cos\theta_m = m\lambda$. To determine the angles of incidence θ_m, we need to first calculate λ. The wavelength is related to the photon's energy as $E = hc/\lambda$. Thus,

$$\lambda = \frac{hc}{E} = \frac{(6.63\times10^{-34}\text{ Js})(3.0\times10^{8}\text{ m/s})}{1.50\times10^{-15}\text{ J}} = 1.326\times10^{-10}\text{ m}$$

From the Bragg condition,

$$\theta_m = \cos^{-1}\left(\frac{m\lambda}{2d}\right) = \cos^{-1}\left[\frac{(1.326\times10^{-10}\text{ m})m}{2(0.21\times10^{-9}\text{ m})}\right] = \cos^{-1}(0.3157m) \Rightarrow \theta_1 = \cos^{-1}(0.3157) = 71.6°$$

Likewise, $\theta_2 = \cos^{-1}(0.3157\times2) = 50.8°$ and $\theta_3 = 18.7°$. Note that $\theta_4 = \cos^{-1}(0.3157\times4)$ is not allowed because the $\cos\theta$ cannot be larger than 1. Thus, the x-rays will be diffracted at angles of incidence equal to 18.7°, 50.8°, and 71.6°.

25.29. Model: Particles have a de Broglie wavelength given by $\lambda = h/p$. The wave nature of the particles causes an interference pattern in a double-slit apparatus.
Solve: **(a)** Since the speed of the neutron and electron are the same, the neutron's momentum is

$$p_\text{n} = m_\text{n}v_\text{n} = \frac{m_\text{n}}{m_\text{e}}m_\text{e}v_\text{n} = \frac{m_\text{n}}{m_\text{e}}m_\text{e}v_\text{e} = \frac{m_\text{n}}{m_\text{e}}p_\text{e}$$

where m_n and m_e are the neutron's and electron's masses. The de Broglie wavelength for the neutron is

$$\lambda_\text{n} = \frac{h}{p_\text{n}} = \frac{h}{p_\text{e}}\frac{p_\text{e}}{p_\text{n}} = \lambda_\text{e}\frac{m_\text{e}}{m_\text{n}}$$

From Section 22.2 on double-slit interference, the fringe spacing is $\Delta y = \lambda L/d$. Thus, the fringe spacing for the electron and neutron are related by

$$\Delta y_\text{n} = \frac{\lambda_\text{n}}{\lambda_\text{e}}\Delta y_\text{e} = \frac{m_\text{e}}{m_\text{n}}\Delta y_\text{e} = \left(\frac{9.11\times10^{-31}\text{ kg}}{1.67\times10^{-27}\text{ kg}}\right)(1.5\times10^{-3}\text{ m}) = 8.18\times10^{-7}\text{ m} = 0.818\ \mu\text{m}$$

(b) If the fringe spacing has to be the same for the neutrons and the electrons,

$$\Delta y_\text{e} = \Delta y_\text{n} \Rightarrow \lambda_\text{e} = \lambda_\text{n} \Rightarrow \frac{h}{m_\text{e}v_\text{e}} = \frac{h}{m_\text{n}v_\text{n}} \Rightarrow v_\text{n} = v_\text{e}\frac{m_\text{e}}{m_\text{n}} = (2.0\times10^{6}\text{ m/s})\left(\frac{9.11\times10^{-31}\text{ kg}}{1.67\times10^{-27}\text{ kg}}\right) = 1.1\times10^{3}\text{ m/s}$$

25.31. Model: Neutrons have a de Broglie wavelength given by $\lambda = h/p$. The wave nature of the neutrons causes a double-slit interference pattern.
Solve: Measurements show that the spacing between the $m = 1$ and $m = -1$ peaks is 1.4 times as long as the length of the reference bar, which gives the real fringe separation $\Delta y = 70\ \mu$m. Similarly, the spacing between the $m = 2$ and $m = -2$ is 2.8 times as long as the length of the reference bar and yields $\Delta y = 70\ \mu$m.
The fringe separation in a double-slit experiment is $\Delta y = \lambda L/d$. Hence,

$$\lambda = \frac{\Delta y\, d}{L} \Rightarrow \frac{h}{p} = \frac{h}{mv} = \frac{\Delta y\, d}{L} \Rightarrow v = \frac{hL}{\Delta y\, md} = \frac{(6.63\times10^{-34}\text{ Js})(3.0\text{ m})}{(70\times10^{-6}\text{ m})(1.67\times10^{-27}\text{ kg})(0.10\times10^{-3}\text{ m})} = 170\text{ m/s}$$

25.39. Model: A particle confined in a one-dimensional box of length L has the discrete energy levels given by Equation 24.14.
Solve: **(a)** Since the energy is entirely kinetic energy,

$$E_n = \frac{h^2}{8mL^2}n^2 = \frac{p^2}{2m} = \tfrac{1}{2}mv_n^2 \Rightarrow v_n = \frac{h}{2mL}n \qquad n = 1, 2, 3, \ldots$$

(b) The first allowed velocity is

$$v_1 = \frac{6.63\times10^{-34}\text{Js}}{2\left(9.11\times10^{-31}\text{ kg}\right)\left(0.20\times10^{-9}\text{ m}\right)} = 1.82\times10^{6}\text{ m/s}$$

For $n = 2$ and $n = 3$, $v_2 = 3.64 \times 10^6$ m/s and $v_3 = 5.46 \times 10^6$ m/s.

26

ELECTRIC CHARGES AND FORCES

Exercises and Problems

26.3. Model: Use the charge model and the model of a conductor as material through which electrons move.
Solve: **(a)** The charge of a plastic rod decreases from −15 nC to −10 nC. That is, −5 nC charge has been removed from the plastic. Because it is the negatively charged electrons that are transferred, −5 nC has been added to the metal sphere.
(b) Because each electron has a charge of 1.60×10^{-19} C and a charge of 5 nC was transferred, the number of electrons transferred from the plastic rod to the metal sphere is

$$\frac{5\times10^{-9}\text{ C}}{1.60\times10^{-19}\text{ C}} = 3.1\times10^{10}$$

26.7. Model: Use the charge model and the model of a conductor as a material through which electrons move.
Visualize:

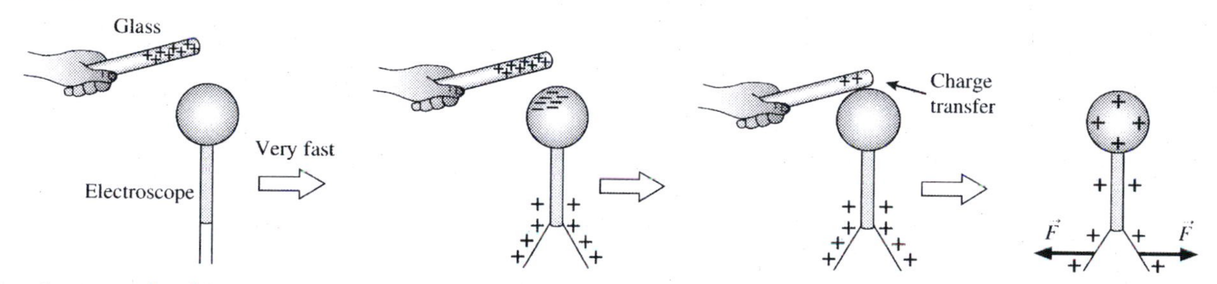

The charge carriers in a metal electroscope are the negative electrons. As the positive rod is brought near, electrons are attracted toward it and move to the top of the electroscope. The electroscope leaves now have a net positive charge, due to the missing electrons, and thus repel each other. At this point, the electroscope as a whole is still neutral (no *net* charge) but has been polarized. On contact, some of the electrons move to the positive rod to neutralize some (but not necessarily all) of the rod's positive charge. After contact, the electroscope *does* have a net positive charge. When the rod is removed, the net positive charge on the electroscope quickly spreads to cover the entire electroscope. The net positive charge on the leaves causes them to continue to repel.

26.13. Model: Model the charged masses as point charges.
Visualize:

Solve: **(a)** The charge q_1 exerts a force $\vec{F}_{1\text{ on }2}$ on q_2 to the right, and the charge q_2 exerts a force $\vec{F}_{2\text{ on }1}$ on q_1 to the left. Using Coulomb's law,

$$F_{1\text{ on }2} = F_{2\text{ on }1} = \frac{K|q_1||q_2|}{r_{12}^2} = \frac{(9.0\times10^9\text{ N m}^2/\text{C}^2)(10\times10^{-6}\text{ C})(10\times10^{-6}\text{ C})}{(1.0\text{ m})^2} = 0.90\text{ N}$$

(b) Newton's second law on either q_1 or q_2 is

$$F_{1\text{ on }2} = m_1 a_1 \Rightarrow a_1 = \frac{0.90\text{ N}}{1.0\text{ kg}} = 0.90\text{ m/s}^2$$

Assess: Even a micro-Couomb is a lot of charge. That is why $F_{1\text{ on }2}$ (or $F_{2\text{ on }1}$) is a measurable force.

26.15. Model: Model the glass bead and the ball bearing as point charges.
Visualize:

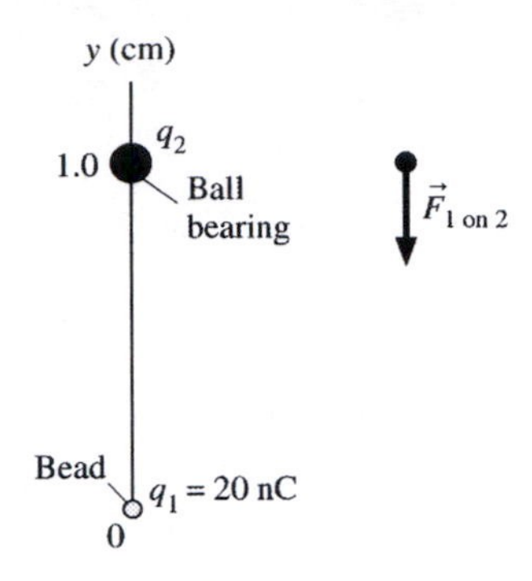

The ball bearing experiences a downward electric force $\vec{F}_{1\text{ on }2}$. By Newton's third law, $F_{2\text{ on }1} = F_{1\text{ on }2}$.
Solve: Using Coulomb's law,

$$F_{1\text{ on }2} = K\frac{|q_1||q_2|}{r_{12}^2} \Rightarrow 0.018\text{ N} = \frac{(9.0\times10^9\text{ N m}^2/\text{C}^2)(20\times10^{-9}\text{ C})|q_2|}{(1.0\times10^{-2}\text{ m})^2} \Rightarrow |q_2| = 1.0\times10^{-8}\text{ C}$$

Because the force $F_{1\text{ on }2}$ is attractive and q_1 is a positive charge, the charge q_2 is a negative charge. Thus, $q_2 = -1.0\times10^{-8}\text{ C} = -10\text{ nC}$.

26.25. Model: A field is the agent that exerts an electric force on a charge.
Visualize:

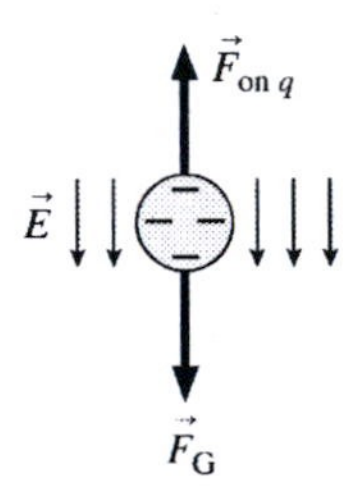

Solve: Newton's second law on the plastic ball is $\Sigma(\vec{F}_{\text{net}})_y = \vec{F}_{\text{on }q} - \vec{F}_G$. To balance the gravitational force with the electric force,

$$F_{\text{on q}} = F_G \Rightarrow |q|E = mg \Rightarrow E = \frac{mg}{|q|} = \frac{(1.0\times10^{-3}\text{ kg})(9.8\text{ N/kg})}{3.0\times10^{-9}\text{ C}} = 3.3\times10^6\text{ N/C}$$

Because $F_{\text{on }q}$ must be upward and the charge is negative, the electric field at the location of the plastic ball must be pointing downward. Thus $\vec{E} = (3.3\times10^6\text{ N/C, downward})$.

Assess: $\vec{F} = q\vec{E}$ means the sign of the charge q determines the direction of $\vec{F}$ or $\vec{E}$. For positive q, $\vec{E}$ and $\vec{F}$ are pointing in the same direction. But $\vec{E}$ and $\vec{F}$ point in opposite directions when q is negative.

26.27. Model: The electric field is that of a positive point charge located at the origin.
Visualize:

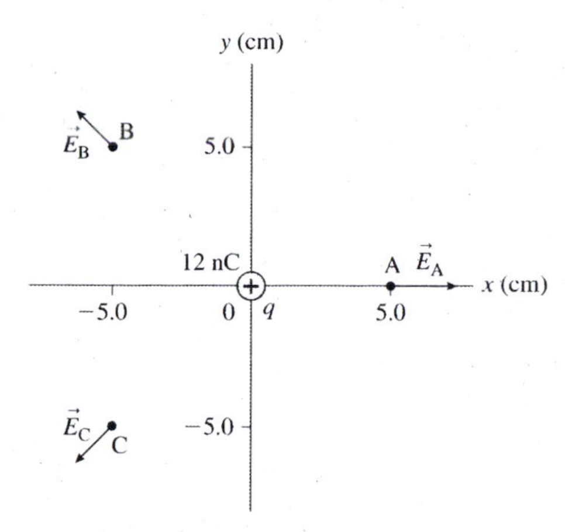

The positions (5.0 cm, 0.0 cm), (−5.0 cm, 5.0 cm), and (−5.0 cm, −5.0 cm) are denoted by A, B, and C, respectively.
Solve: **(a)** The electric field for a positive charge is

$$\vec{E} = \left(\frac{1}{4\pi\varepsilon_0}\frac{q}{r^2}, \text{ away from } q\right)$$

Using $1/4\pi\varepsilon_0 = 9.0\times10^9 \text{ N m}^2/\text{C}^2$ and $q = 12\times10^{-9}$ C,

$$\vec{E} = \left(\frac{108 \text{ N m}^2/\text{C}}{r^2}, \text{ away from } q\right)$$

The electric fields at points A, B, and C are

$$\vec{E}_A = \frac{108 \text{ N m}^2/\text{C}}{\left(5.0\times10^{-2} \text{ m}\right)^2}\hat{i} = 4.3\times10^4\hat{i} \text{ N/C}$$

$$\vec{E}_B = \frac{108 \text{ N m}^2/\text{C}}{\left(-5.0\times10^{-2} \text{ m}\right)^2 + \left(5.0\times10^{-2} \text{ m}\right)^2}\left[\frac{1}{\sqrt{2}}\left(-\hat{i}+\hat{j}\right)\right] = \left(-1.53\times10^4\hat{i} + 1.53\times10^4\hat{j}\right) \text{ N/C}$$

$$\vec{E}_C = \frac{108 \text{ N m}^2/\text{C}}{\left(-5.0\times10^{-2} \text{ m}\right)^2 + \left(-5.0\times10^{-2} \text{ m}\right)^2}\left[\frac{1}{\sqrt{2}}\left(-\hat{i}-\hat{j}\right)\right] = \left(-1.53\times10^4\hat{i} - 1.53\times10^4\hat{j}\right) \text{ N/C}$$

(b) The three vectors are shown in the diagram.
Assess: The vectors $\vec{E}_A$, $\vec{E}_B$, and $\vec{E}_C$ are pointing away from the positive charge.

26.33. Model: The protons are point charges.
Solve: **(a)** The electric force between the protons is

$$F_E = K\frac{|q_1||q_2|}{r^2} = \frac{\left(9.0\times10^9 \text{ N m}^2/\text{C}^2\right)\left(1.60\times10^{-19} \text{ C}\right)\left(1.60\times10^{-19} \text{ C}\right)}{\left(2.0\times10^{-15} \text{ m}\right)^2} = 58 \text{ N}$$

(b) The gravitational force between the protons is

$$F_G = \frac{G\,m_1m_2}{r^2} = \frac{\left(6.67\times10^{-11} \text{ N m}^2/\text{kg}^2\right)\left(1.67\times10^{-27} \text{ kg}\right)\left(1.67\times10^{-27} \text{ kg}\right)}{\left(2.0\times10^{-15} \text{ m}\right)^2} = 4.7\times10^{-35} \text{ N}$$

(c) The ratio of the electric force to the gravitational force is

$$\frac{F_E}{F_G} = \frac{58 \text{ N}}{4.7\times10^{-35} \text{ N}} = 1.23\times10^{36}$$

26.37. Model: The charges are point charges.
Visualize: Please refer to Figure P26.37.
Solve: The electric force on charge q_1 is the vector sum of the forces $\vec{F}_{2\text{ on }1}$ and $\vec{F}_{3\text{ on }1}$, where q_1 is the 1.0 nC charge, q_2 is the 2.0 nC charge, and q_3 is the other 2.0 nC charge. We have

$$\vec{F}_{2\text{ on }1} = \left(\frac{K|q_1||q_2|}{r^2}, \text{ away from } q_2\right)$$
$$= \left(\frac{(9.0\times10^9 \text{ N m}^2/\text{C}^2)(1.0\times10^{-9}\text{ C})(2.0\times10^{-9}\text{ C})}{(1.0\times10^{-2}\text{ m})^2}, \text{ away from } q_2\right)$$
$$= (1.8\times10^{-4}\text{ N, away from } q_2) = (1.8\times10^{-4}\text{ N})(\cos60°\hat{\imath} + \sin60°\hat{\jmath})$$

$$\vec{F}_{3\text{ on }1} = \left(\frac{K|q_1||q_3|}{r_2}, \text{ away from } q_3\right) = (1.8\times10^{-4}\text{ N, away from } q_3) = (1.8\times10^{-4}\text{ N})(-\cos60°\hat{\imath} + \sin60\hat{\jmath})$$

$$\Rightarrow \vec{F}_{\text{on }1} = \vec{F}_{2\text{ on }1} + \vec{F}_{3\text{ on }1} = 2(1.8\times10^{-4}\text{ N})\sin60°\hat{\jmath} = 3.1\times10^{-4}\hat{\jmath}\text{ N}$$

So the force on the 1.0 nC charge is 3.1×10^{-4} N directed upward.

26.41. Model: The charges are point charges.
Visualize:

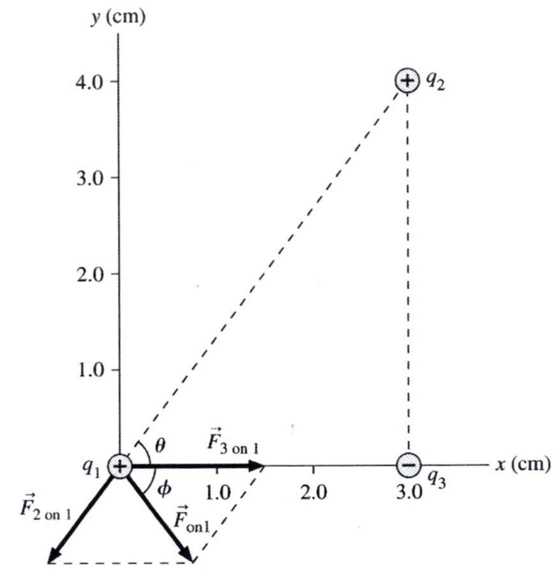

Solve: The electric force on charge q_1 is the vector sum of the forces $\vec{F}_{2\text{ on }1}$ and $\vec{F}_{3\text{ on }1}$. We have

$$\vec{F}_{2\text{ on }1} = \left(\frac{K|q_1||q_2|}{r^2}, \text{ away from } q_2\right)$$
$$= \left(\frac{(9.0\times10^9 \text{ N m}^2/\text{C}^2)(5\times10^{-9}\text{ C})(10\times10^{-9}\text{ C})}{(4.0\times10^{-2}\text{ m})^2 + (3.0\times10^{-2}\text{ m})^2}, \text{ away from } q_2\right)$$
$$= (1.8\times10^{-4}\text{ N, away from } q_2) = (1.8\times10^{-4}\text{ N})(-\cos\theta\hat{\imath} - \sin\theta\hat{\jmath})$$

From the geometry of the figure,

$$\tan\theta = \frac{4.0 \text{ cm}}{3.0 \text{ cm}} \Rightarrow \theta = 53.13° \Rightarrow \vec{F}_{2 \text{ on } 1} = \left(1.8\times10^{-4} \text{ N}\right)\left(-0.6\hat{\imath} - 0.8\hat{\jmath}\right)$$

$$\vec{F}_{3 \text{ on } 1} = \left(\frac{\left(9.0\times10^{9} \text{ N m}^2/\text{C}^2\right)\left(5\times10^{-9} \text{ C}\right)\left(5\times10^{-9} \text{ C}\right)}{\left(3.0\times10^{-2} \text{ m}\right)^2}, \text{ toward } q_3\right) = \left(2.5\times10^{-4} \text{ N, toward } q_3\right) = 2.5\times10^{-4}\hat{\imath} \text{ N}$$

$$\Rightarrow \vec{F}_{\text{on } 1} = \vec{F}_{2 \text{ on } 1} + \vec{F}_{3 \text{ on } 1} = \left(1.42\times10^{-4}\hat{\imath} - 1.44\times10^{-4}\hat{\jmath}\right) \text{ N}$$

The magnitude and direction of the resultant force vector are

$$F_{\text{on } 1} = \sqrt{\left(1.42\times10^{-4} \text{ N}\right)^2 + \left(-1.44\times10^{-4} \text{ N}\right)^2} = 2.0\times10^{-4} \text{ N}$$

$$\phi = \tan^{-1}\left(\frac{1.44\times10^{-4} \text{ N}}{1.42\times10^{-4} \text{ N}}\right) = 45° \text{ clockwise from the } +x\text{-axis.}$$

26.45. Model: The charges are point charges.
Visualize: Please refer to Figure P26.45.
Solve: Placing the 1.0 nC charge at the origin and calling it q_1, the –6.0 nC is q_3, the q_2 charge is in the first quadrant, and the q_4 charge is in the second quadrant. The net electric force on q_1 is the vector sum of the electric forces from the other three charges q_2, q_3, and q_4. We have

$$\vec{F}_{2 \text{ on } 1} = \left(\frac{K|q_1||q_2|}{r^2}, \text{ away from } q_2\right)$$

$$= \left(\frac{\left(9.0\times10^{9} \text{ N m}^2/\text{C}^2\right)\left(1.0\times10^{-9} \text{ C}\right)\left(2.0\times10^{-9} \text{ C}\right)}{\left(5.0\times10^{-2} \text{ m}\right)^2}, \text{ away from } q_2\right)$$

$$= \left(0.72\times10^{-5} \text{ N, away from } q_2\right) = \left(0.72\times10^{-5} \text{ N}\right)\left(-\cos45°\hat{\imath} - \sin45°\hat{\jmath}\right)$$

$$\vec{F}_{3 \text{ on } 1} = \left(\frac{K|q_1||q_3|}{r^2}, \text{ toward } q_3\right)$$

$$= \left(\frac{\left(9.0\times10^{9} \text{ N m}^2/\text{C}^2\right)\left(1.0\times10^{-9} \text{ C}\right)\left(6.0\times10^{-9} \text{ C}\right)}{\left(5.0\times10^{-2} \text{ m}\right)^2}, \text{ toward } q_3\right)$$

$$= \left(2.16\times10^{-5} \text{ N, away from } q_3\right) = 2.16\times10^{-5}\hat{\jmath} \text{ N}$$

$$\vec{F}_{4 \text{ on } 1} = \left(\frac{K|q_1||q_4|}{r^2}, \text{ away from } q_4\right) = \left(0.72\times10^{-5} \text{ N}\right)\left(\cos45°\hat{\imath} - \sin45°\hat{\jmath}\right)$$

$$\Rightarrow \vec{F}_{\text{on } 1} = \vec{F}_{2 \text{ on } 1} + \vec{F}_{3 \text{ on } 1} + \vec{F}_{4 \text{ on } 1} = \left[\left(2.16\times10^{-5} \text{ N}\right) - \left(0.72\times10^{-5} \text{ N}\right)\left(2\sin45°\right)\right]\hat{\jmath} = 1.14\times10^{-5}\hat{\jmath} \text{ N}$$

26.49. Model: The charged particles are point charges.
Visualize: Please refer to Figure P26.49.
Solve: The charge q_2 is in static equilibrium, so the net electric field at the location of q_2 is zero. We have

$$\vec{E}_{\text{net}} = \vec{E}_{q_1} + \vec{E}_{-3 \text{ nC}} = \frac{1}{4\pi\varepsilon_0}\frac{|q_1|}{(0.20 \text{ m})^2}\left(\pm\hat{\imath}\right) + \frac{1}{4\pi\varepsilon_0}\frac{\left(3.0\times10^{-9} \text{ C}\right)}{(0.10 \text{ m})^2}\left(-\hat{\imath}\right) = 0 \text{ N/C}$$

We have used the $\pm$ sign to indicate that a positive charge on q_1 leads to an electric field along $+\hat{\imath}$ and a negative charge on q_1 leads to an electric field along $-\hat{\imath}$. Because the above equation can only be satisfied if we use $+\hat{\imath}$, we infer that the charge q_1 is a positive charge. Thus,

$$\frac{q_1}{(0.20 \text{ m})^2}\left(+\hat{\imath}\right) - \frac{3.0\times10^{-9} \text{ C}}{(0.10 \text{ m})^2}\left(\hat{\imath}\right) = 0 \text{ N/C} \Rightarrow q_1 = 12.0 \text{ nC}$$

26.51. Model: The charged particles are point charges.

Visualize:

$|x| < a$:

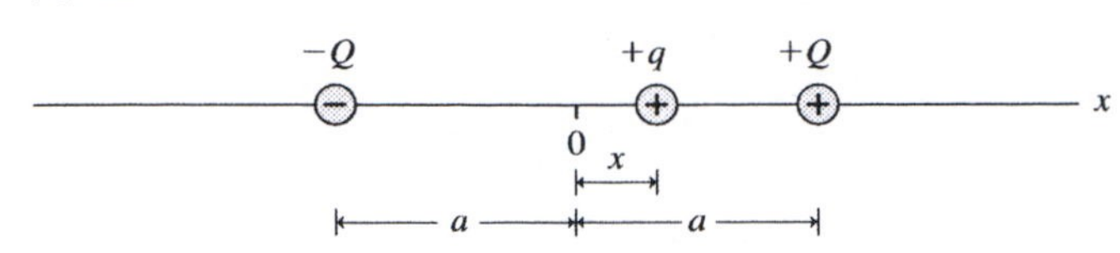

$x > a$:

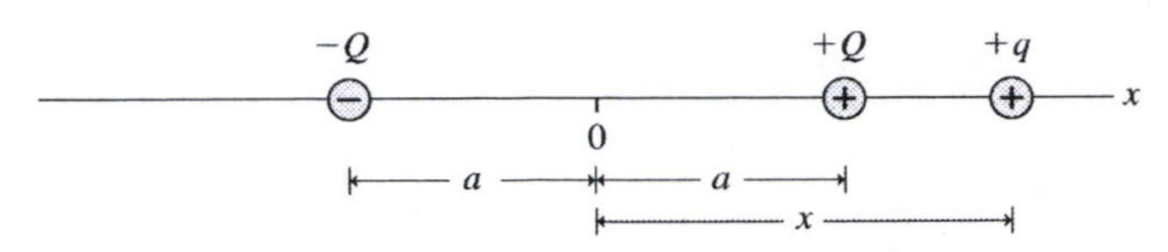

$x < -a$:

+q −Q +Q x

0

a a

x

Solve: **(a)** The force on q is the vector sum of the force from $-Q$ and $+Q$. We have

$$\vec{F}_{+Q \text{ on } +q} = \left(\frac{K|+Q||+q|}{(a-x)^2}, \text{ away from } +Q\right) = \frac{KQq}{(a-x)^2}(-\hat{i})$$

$$\vec{F}_{-Q \text{ on } +q} = \left(\frac{K|-Q||+q|}{(a+x)^2}, \text{ toward } -Q\right) = \frac{KQq}{(a+x)^2}(-\hat{i})$$

$$\Rightarrow (F_{\text{net}})_x = -KQq\left[\frac{1}{(a-x)^2} + \frac{1}{(a+x)^2}\right] = -\frac{2KQq(a^2+x^2)}{(a^2-x^2)^2}$$

To arrive at the final expression we used $(a-x)^2(a+x)^2 = [(a-x)(a+x)]^2 = (a^2-x^2)^2$.

(b) There are two cases when $|x| > a$. For $x > a$,

$$\vec{F}_{+Q \text{ on } +q} = \left(\frac{K|+Q||+q|}{(x-a)^2}, \text{ away from } +Q\right) = \frac{KQq}{(x-a)^2}(+\hat{i}) \qquad \vec{F}_{-Q \text{ on } +q} = \left(\frac{K|-Q||+q|}{(x+a)^2}, \text{ toward } -Q\right) = \frac{KQq}{(x+a)^2}(-\hat{i})$$

$$\Rightarrow (F_{\text{net}})_x = KQq\left[\frac{1}{(x-a)^2} - \frac{1}{(x+a)^2}\right] = \frac{4KQqax}{(x^2-a^2)^2}$$

For $x < -a$ (that is, for negative values of x),

$$\vec{F}_{+Q \text{ on } +q} = \frac{KQq}{(x-a)^2}(-\hat{i}) \qquad \vec{F}_{-Q \text{ on } +q} = \frac{KQq}{(a+x)^2}(+\hat{i})$$

$$\Rightarrow (F_{\text{net}})_x = -\frac{4KQqax}{(x^2-a^2)^2}$$

That is, the net force is to the right when $x > a$ and to the right when $x < -a$. We can combine these two cases into a single equation for $|x| > a$:

$$(F_{\text{net}})_x = \frac{4KQqa|x|}{(x^2-a^2)^2}$$

Here, the force is always to the right when $|x| > a$.

26.55. Model: The electron and the proton are point charges.
Solve: The electric Coulomb force between the electron and the proton provides the centripetal acceleration for the electron's circular motion. Thus,

$$\frac{K(e)(e)}{r^2}=\frac{mv^2}{r}=mr\omega^2$$

$$\Rightarrow f=\sqrt{\frac{Ke^2}{mr^3}}=\sqrt{\frac{(9.0\times10^9\ \text{N m}^2/\text{C}^2)(1.60\times10^{-19}\ \text{C})^2}{(9.11\times10^{-31}\ \text{kg})(5.3\times10^{-11})^3}}=4.12\times10^{16}\ \text{rad/s}\times\frac{1\ \text{rev}}{2\pi\ \text{rad}}=6.6\times10^{15}\ \text{rev/s}$$

26.59. Model: The charged spheres are point charges.
Visualize:

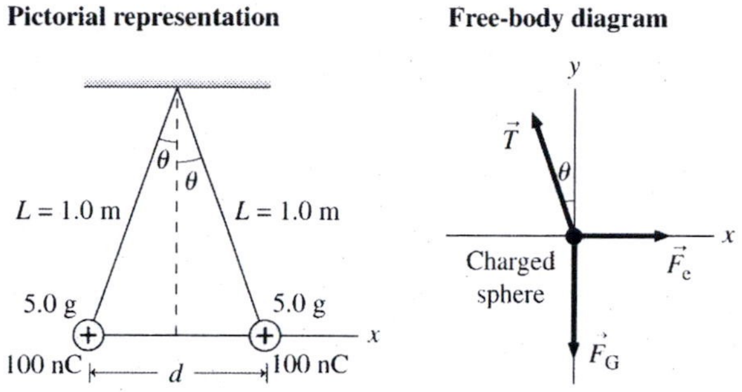

Pictorial representation **Free-body diagram**

Each sphere is in static equilibrium and the string makes an angle θ with the vertical. The three forces acting on each sphere are the electric force, the gravitational force on the sphere, and the tension force.
Solve: In static equilibrium, Newton's first law is $\vec{F}_{\text{net}}=\vec{T}+\vec{F}_G+\vec{F}_e=\vec{0}$. In component form,

$$(F_{\text{net}})_x=T_x+(F_G)_x+(F_e)_x=0\ \text{N} \qquad (F_{\text{net}})=T_y+(F_G)_y+(F_e)_y=0\ \text{N}$$

$$\Rightarrow -T\sin\theta+0\ \text{N}+\frac{Kq^2}{d^2}=0\ \text{N} \qquad T\cos\theta-mg+0\ \text{N}=0\ \text{N}$$

$$\Rightarrow T\sin\theta=\frac{Kq^2}{d^2}=\frac{Kq^2}{(2L\sin\theta)^2} \qquad T\cos\theta=+mg$$

Dividing the two equations,

$$\sin^2\theta\tan\theta=\frac{Kq^2}{4L^2mg}=\frac{(9.0\times10^9\ \text{N m}^2/\text{C}^2)(100\times10^{-9}\ \text{C})^2}{4(1.0\ \text{m})^2(5.0\times10^{-3}\ \text{kg})(9.8\ \text{N/kg})}=4.59\times10^{-4}$$

For small-angles, $\tan\theta\approx\sin\theta$. With this approximation we obtain $\sin\theta=0.07714$ rad and $\theta=4.4°$.

26.63. Model: The electric field is that of a positive point charge located at the origin.
Visualize: Please refer to Figure P26.63. Place the 5.0 nC charge at the origin.
Solve: The electric field is

$$\vec{E}=\left(\frac{1}{4\pi\varepsilon_0}\frac{q}{r^2},\ \text{away from } q\right)=\left(\frac{(9\times10^9\ \text{N m}^2/\text{C}^2)(5.0\times10^{-9}\ \text{C})}{r^2},\ \text{away from } q\right)$$

$$=\left(\frac{45\ \text{N m}^2/\text{C}}{r^2},\ \text{away from } q\right)$$

At each of the three points,

$$\vec{E}_1 = \left(\frac{45 \text{ N m}^2/\text{C}}{(2.0\times10^{-2}\text{ m})^2 + (1.0\times10^{-2}\text{ m})^2}, \text{ away from } q\right) = (9.0\times10^4 \text{ N/C})(\cos\theta\hat{i} + \sin\theta\hat{j})$$

$$= (9.0\times10^4 \text{ N/C})\left(\frac{1}{\sqrt{5}}\hat{i} + \frac{2}{\sqrt{5}}\hat{j}\right) = (4.0\times10^4\hat{i} + 8.0\times10^4\hat{j}) \text{ N/C}$$

$$\vec{E}_2 = \left(\frac{45 \text{ N m}^2/\text{C}}{(1.0\times10^{-2}\text{ m})^2}, \text{ away from } q\right) = 4.5\times10^5\hat{i} \text{ N/C}$$

$$\vec{E}_3 = \left(\frac{45 \text{ N m}^2/\text{C}}{(2.0\times10^{-2}\text{ m})^2 + (1.0\times10^{-2}\text{ m})^2}, \text{ away from } q\right) = (4.0\times10^4\hat{i} - 8.0\times10^4\hat{j}) \text{ N/C}$$

26.65. Model: The electric field is that of a positive charge at $(x, y) = (1.0 \text{ cm}, 2.0 \text{ cm})$.

Visualize:

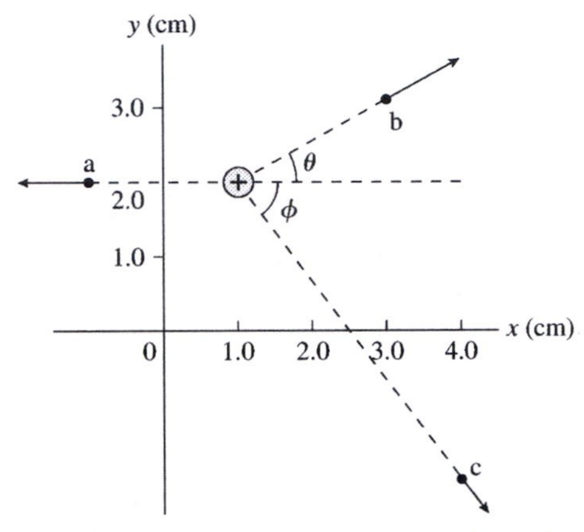

Solve: **(a)** The electric field of a positive charge points straight away from the charge, so we can roughly locate the points of interest based simply on whether the signs of E_x and E_y are positive or negative. For point a, the electric field has no y-component and the x-component points to the left, so point a must be to the left of the charge along a horizontal line. Using the field of a point charge,

$$E_x = E = \frac{K|q|}{r_a^2} \Rightarrow r_a = \sqrt{\frac{K|q|}{|E_x|}} = \sqrt{\frac{(9.0\times10^9 \text{ N m}^2/\text{C}^2)(10.0\times10^{-9}\text{ C})}{225{,}000 \text{ N/C}}} = 0.0200 \text{ m} = 2.00 \text{ cm}$$

Thus, $(x_a, y_a) = (-1 \text{ cm}, 2 \text{ cm})$.

(b) Point b is above and to the right of the charge. The magnitude of the field at this point is

$$E = \sqrt{E_x^2 + E_y^2} = \sqrt{(161{,}000 \text{ N/C})^2 + (80{,}500 \text{ N/C})^2} = 180{,}000 \text{ N/C}$$

Using the field of a point charge,

$$E = \frac{K|q|}{r_b^2} \Rightarrow r_b = \sqrt{\frac{K|q|}{|\vec{E}|}} = \sqrt{\frac{(9.0\times10^9 \text{ N m}^2/\text{C}^2)(10.0\times10^{-9}\text{ C})}{180{,}000 \text{ N/C}}} = 2.236 \text{ cm}$$

This gives the total distance but not the horizontal and vertical components. However, we can determine the angle θ because $\vec{E}_b$ points straight away from the positive charge. Thus,

$$\theta = \tan^{-1}\left(\frac{|E_y|}{|E_x|}\right) = \tan^{-1}\left(\frac{80{,}500 \text{ N/C}}{161{,}000 \text{ N/C}}\right) = \tan^{-1}\left(\frac{1}{2}\right) = 26.57°$$

The horizontal and vertical distances are then $d_x = r_b\cos\theta = (2.236 \text{ cm})\cos 26.57° = 2.00 \text{ cm}$ and $d_y = r_b\sin\theta = 1.00 \text{ cm}$. Thus, point b is at position $(x_b, y_b) = (3 \text{ cm}, 3 \text{ cm})$.

(c) To calculate point c, which is below and to the right of the charge, a similar procedure is followed. We first find $E = 36{,}000\ \text{N/C}$ from which we find the total distance $r_c = 5.00$ cm. The angle ϕ is

$$\phi = \tan^{-1}\left(\frac{|E_y|}{|E_x|}\right) = \tan^{-1}\left(\frac{28{,}800}{21{,}600}\right) = 53.13°$$

which gives distances $d_x = r_c \cos\phi = 3.00$ cm and $d_y = r_c \sin\phi = 4.00$ cm. Thus, point c is at position (x_c, y_c) = (4 cm, -2 cm).

26.67. Model: The charged ball attached to the string is a point charge.
Visualize:

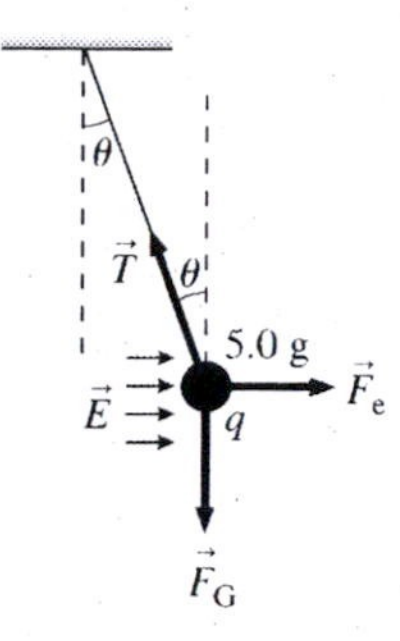

The ball is in static equilibrium in the external electric field when the string makes an angle $\theta = 20°$ with the vertical. The three forces acting on the charged ball are the electric force due to the field, the gravitational force on the ball, and the tension force.

Solve: In static equilibrium, Newton's second law for the ball is $\vec{F}_{net} = \vec{T} + \vec{F}_G + \vec{F}_e = \vec{0}$. In component form,

$$(F_{net})_x = T_x + 0\ \text{N} + qE = 0\ \text{N} \qquad (F_{net})_y = T_y - mg + 0\ \text{N} = 0\ \text{N}$$

The above two equations simplify to

$$T\sin\theta = qE \qquad T\cos\theta = mg$$

Dividing both equations, we get

$$\tan\theta = \frac{qE}{mg} \Rightarrow q = \frac{mg\tan\theta}{E} = \frac{(5.0\times10^{-3}\ \text{kg})(9.8\ \text{N/kg})\tan 20°}{100{,}000\ \text{N/C}} = 1.78\times10^{-7}\ \text{C} = 178\ \text{nC}$$

26.71. Solve: **(a)** At what distance from a 15 nC charge is the electric field strength 54,000 N/C?
(b) The distance is

$$r = \sqrt{\frac{(9.0\times10^{9}\ \text{N m}^2/\text{C}^2)(15\times10^{-9}\ \text{C})}{54{,}000\ \text{N/C}}} = 0.050\ \text{m} = 5.0\ \text{cm}$$

THE ELECTRIC FIELD

27

Exercises and Problems

27.1. Model: The electric field is the superposition of the fields of two point charges. We must add vectors.
Visualize: The fields of both charges are straight away from the charges.

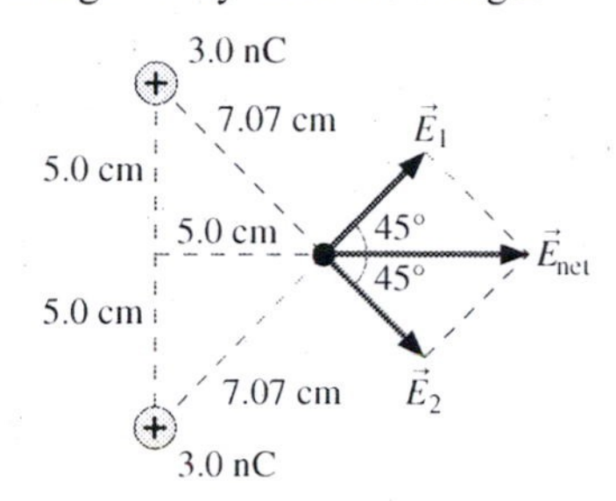

Solve: At the point of interest, the electric field $\vec{E}_1$ of the top charge is

$$\vec{E}_1 = \left(\frac{q_1}{4\pi\varepsilon_0 r_1^2}, 45° \text{ below } x\text{-axis}\right) = \frac{(9.0\times10^9 \text{ N m}^2/\text{C}^2)(3.0\times10^{-9}\text{ C})}{(0.0707\text{ m})^2}(\cos 45°\hat{i} - \sin 45°\hat{j})$$

$$= (3820\,\hat{i} - 3820\,\hat{j}) \text{ N/C}$$

Similarly, field $\vec{E}_2$ of the lower charge is

$$\vec{E}_2 = \left(\frac{q_2}{4\pi\varepsilon_0 r_1^2}, 45° \text{ above } x\text{-axis}\right) = (3820\,\hat{i} + 3820\,\hat{j}) \text{ N/C}$$

The net electric field of the two charges is

$$\vec{E}_{net} = \vec{E}_1 + \vec{E}_2 = 7600\,\hat{i} \text{ N/C} = (7600 \text{ N/C, horizontal})$$

Thus, the strength of the electric field is 7.6×10^3 N/C and its direction is horizontal.
Assess: The symmetry of the situation causes the y-components of the two fields to cancel. We were able to see this from the figure.

27.5. Model: The distances to the observation points are large compared to the size of the dipole, so model the field as that of a dipole moment.

Visualize:

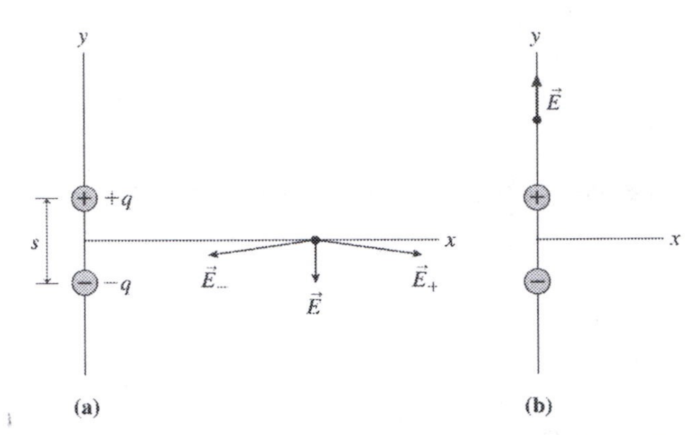

The dipole consists of charges $\pm q$ along the y-axis. The electric field in (a) points down. The field in (b) points up.

Solve: **(a)** The dipole moment is

$$\vec{p} = (qs,\ \text{from} - \text{to} +) = (1.0\times10^{-9}\ \text{C})(0.0020\ \text{m})\hat{j} = 2.0\times10^{-12}\,\hat{j}\ \text{C m}$$

The electric field at (10 cm, 0 cm), which is at distance $r = 0.10$ m in the plane perpendicular to the electric dipole, is

$$\vec{E} = -\frac{1}{4\pi\varepsilon_0}\frac{\vec{p}}{r^3} = -(9.0\times10^9\ \text{N m}^2/\text{C}^2)\frac{2.0\times10^{-12}\,\hat{j}\ \text{C m}}{(0.10\ \text{m})^3} = -18.0\hat{j}\ \text{N/C}$$

The field strength, which is all we're asked for, is 18.0 N/C.

(b) The electric field at (0 cm, 10 cm), which is at $r = 0.10$ m along the axis of the dipole, is

$$\vec{E} = \frac{1}{4\pi\varepsilon_0}\frac{2\vec{p}}{r^3} = (9.0\times10^9\ \text{N m}^2/\text{C}^2)\frac{2(2.0\times10^{-12}\,\hat{j}\ \text{C m})}{(0.10\ \text{m})^3} = 36\hat{j}\ \text{N/C}$$

The field strength at this point is 36 N/C.

27.9. Model: The rods are thin. Assume that the charge lies along a *line*.

Visualize:

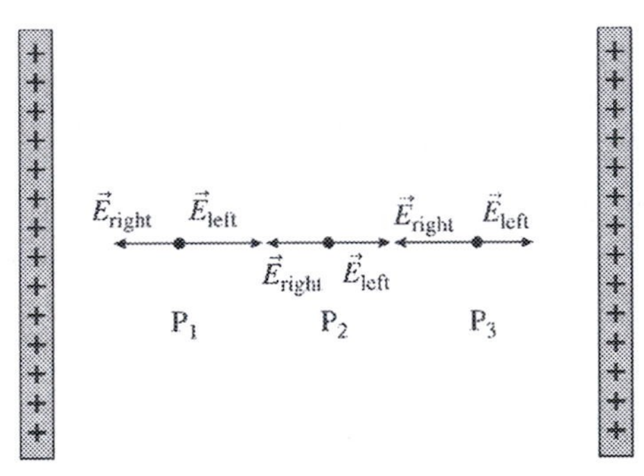

Because both the rods are positively charged, the electric field from each rod points away from the rod. Because the electric fields from the two rods are in opposite directions at P_1, P_2, and P_3, the net field strength at each point is the difference of the field strengths from the two rods.

Solve: Example 27.3 gives the electric field strength in the plane that bisects a charged rod:

$$E_{\text{rod}} = \frac{1}{4\pi\varepsilon_0}\frac{|Q|}{r\sqrt{r^2 + (L/2)^2}}$$

The electric field from the rod on the right at a distance of 1 cm from the rod is

$$E_{\text{right}} = (9.0\times10^9\ \text{N m}^2/\text{C}^2)\frac{10\times10^{-9}\ \text{C}}{(0.01\ \text{m})\sqrt{(0.01\ \text{m})^2 + (0.05\ \text{m})^2}} = 1.765\times10^5\ \text{N/C}$$

The electric field from the rod on the right at distances 2 cm and 3 cm from the rod are 0.835×10^5 N/C and 0.514×10^5 N/C. The electric fields produced by the rod on the left at the same distances are the same. Point P_1 is 1.0 cm from the rod on the left and is 3.0 cm from the rod on the right. Because the electric fields at P_1 have opposite directions, the net electric field strengths are

At 1.0 cm $\quad E = 1.765\times10^5\ \text{N/C} - 0.514\times10^5\ \text{N/C} = 1.25\times10^5\ \text{N/C}$

At 2.0 cm $\quad E = 0.835\times10^5\ \text{N/C} - 0.835\times10^5\ \text{N/C} = 0\ \text{N/C}$

At 3.0 cm $\quad E = 1.765\times10^5\ \text{N/C} - 0.514\times10^5\ \text{N/C} = 1.25\times10^5\ \text{N/C}$

27.13. Model: Each disk is a uniformly charged disk. When the disk is charged negatively, the on-axis electric field of the disk points toward the disk. The electric field points away from the disk for a positively charged disk.
Visualize:

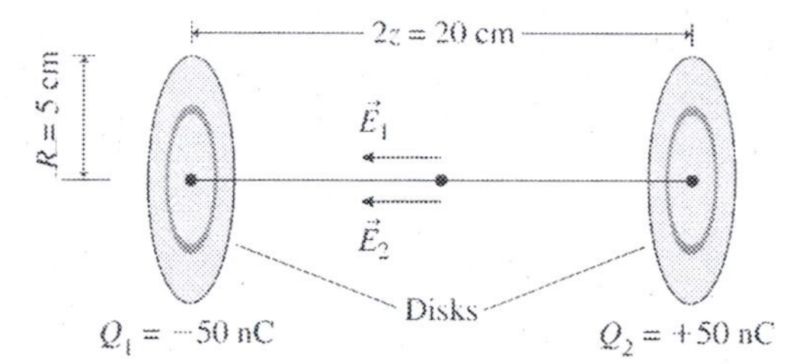

Solve: **(a)** The surface charge density on the disk is

$$|\eta| = \frac{|Q|}{A} = \frac{|Q|}{\pi R^2} = \frac{50\times10^{-9}\ \text{C}}{\pi(0.050\ \text{m})^2} = 6.366\times10^{-6}\ \text{C/m}^2$$

From Equation 27.23, the electric field of the left disk at $z = 0.10$ m is

$$(E_1)_z = \frac{\eta}{2\varepsilon_0}\left[1 - \frac{1}{\sqrt{1+R^2/z^2}}\right] = \frac{-6.366\times10^{-6}\ \text{C/m}^2}{2(8.85\times10^{-12}\ \text{C}^2/\text{N m}^2)}\left[1 - \frac{1}{\sqrt{1+(0.050\ \text{m}/0.10\ \text{m})^2}}\right] = -38{,}000\ \text{N/C}$$

In other words, $\vec{E}_1 = (38{,}000\ \text{N/C, left})$. Similarly, the electric field of the right disk at $z = 0.10$ m (to its left) is $\vec{E}_2 = (38{,}000\ \text{N/C, left})$. The net field at the midpoint between the two rings is $\vec{E} = \vec{E}_1 + \vec{E}_2 = (7.6\times10^4\ \text{N/C, left})$.
(b) The force on the charge is

$$\vec{F} = q\vec{E} = (-1.0\times10^{-9}\ \text{C})(7.6\times10^4\ \text{N/C, left}) = (7.6\times10^{-5}\ \text{N, right})$$

Assess: Note that the force on the negative charge is to the right because the electric field is to the left.

27.15. Model: The distance 2.0 mm is very small in comparison to the size of the electrode, so we can model the electrode as a plane of charge.
Solve: From Equation 27.26, the electric field of a plane of charge is

$$E_{\text{plane}} = \frac{\eta}{2\varepsilon_0} = \frac{Q}{2\varepsilon_0 A} = \frac{80\times10^{-9}\ \text{C}}{2(8.85\times10^{-12}\ \text{C}^2/\text{N m}^2)(0.20\times0.20\ \text{m})} = 1.13\times10^5\ \text{N/C}$$

27.17. Model: The electric field is uniform in a region of space between closely spaced capacitor plates.
Solve: The electric field inside a capacitor is $E = Q/\varepsilon_0 A$. Thus, the charge needed to produce a field of strength E is

$$Q = \varepsilon_0 AE = (8.85\times10^{-12}\ \text{C}^2/\text{N m}^2)(0.04\ \text{m}\times0.04\ \text{m})(1.0\times10^6\ \text{N/C}) = 14.2\ \text{nC}$$

Thus, one plate has a charge of 14.2 nC and the other has a charge of -14.2 nC.
Assess: Note that the capacitor as a whole has no net charge.

27.23. Model: The infinite charged plane produces a uniform electric field.
Solve: **(a)** The electric field of a plane of charge with surface charge density η is

$$E = \frac{\eta}{2\varepsilon_0} \Rightarrow \eta = 2\varepsilon_0 E = 2\varepsilon_0\left(\frac{F}{q}\right) = 2\varepsilon_0\left(\frac{ma}{q}\right)$$

where m is the mass, q is the charge, and a is the acceleration of the electron. To obtain η we must first find a. From the kinematic of motion equation $v_1^2 = v_0^2 + 2a\Delta x$,

$$a = \frac{v_1^2 - v_0^2}{2\Delta x} = \frac{(1.0\times10^7 \text{ m/s})^2 - (0 \text{ m/s})^2}{2(2.0\times10^{-2} \text{ m})} = 2.5\times10^{15} \text{ m/s}^2$$

$$\Rightarrow \eta = \frac{2(8.85\times10^{-12} \text{ C}^2/\text{N m}^2)(9.11\times10^{-31} \text{ kg})(2.5\times10^{15} \text{ m/s}^2)}{1.60\times10^{-19} \text{ C}} = 2.5\times10^{-7} \text{ C/m}^2$$

(b) Using the kinematic equation of motion $v_1 = v_0 + a\Delta t$,

$$\Delta t = \frac{v_1 - v_0}{a} = \frac{1.0\times10^7 \text{ m/s} - 0 \text{ m/s}}{2.5\times10^{15} \text{ m/s}^2} = 4.0\times10^{-9} \text{ s}$$

27.29. Model: The electric field is that of three point charges q_1, q_2, and q_3.
Visualize: Please refer to Figure P27.29. Assume the charges are in the x-y plane. The –5.0 nC charge is q_1, the bottom 10 nC charge is q_3, and the top 10 nC charge is q_2. The net electric field at the dot is $\vec{E}_{\text{net}} = \vec{E}_1 + \vec{E}_2 + \vec{E}_3$. The procedure will be to find the magnitudes of the electric fields, to write them in component form, and to add the components.
Solve: (a) The electric field produced by q_1 is

$$E_1 = \frac{1}{4\pi\varepsilon_0}\frac{|q_1|}{r_1^2} = \frac{(9.0\times10^9 \text{ N m}^2/\text{C}^2)(5.0\times10^{-9} \text{ C})}{(0.020 \text{ m})^2} = 112{,}500 \text{ N/C}$$

$\vec{E}_1$ points toward q_1, so in component form $\vec{E}_1 = -112{,}500\hat{j}$ N/C. The electric field produced by q_2 is $E_2 = 56{,}250$ N/C. $\vec{E}_2$ points away from q_2, so $\vec{E}_2 = -56{,}250\hat{i}$ N/C. Finally, the electric field produced by q_3 is

$$E_3 = \frac{1}{4\pi\varepsilon_0}\frac{|q_3|}{r_3^2} = \frac{(9.0\times10^9 \text{ N m}^2/\text{C}^2)(10\times10^{-9} \text{ C})}{(0.020 \text{ m})^2 + (0.040 \text{ m})^2} = 45{,}000 \text{ N/C}$$

$\vec{E}_3$ points away from q_3 and makes an angle $\phi = \tan^{-1}(2/4) = 26.6°$ with the x-axis. So,

$$\vec{E}_3 = -E_3\cos\phi\hat{i} + E_3\sin\phi\hat{j} = (-40{,}250\hat{i} + 20{,}130\hat{j}) \text{ N/C}$$

Adding these three vectors gives

$$\vec{E}_{\text{net}} = \vec{E}_1 + \vec{E}_2 + \vec{E}_3 = (-96{,}500\hat{i} - 92{,}400\hat{j}) \text{ N/C} = (-9.7\times10^4\hat{i} - 9.2\times10^4\hat{j}) \text{ N/C}$$

This is in component form.
(b) The magnitude of the field is

$$E_{\text{net}} = \sqrt{E_x^2 + E_y^2} = \sqrt{(96{,}500 \text{ N/C})^2 + (92{,}400 \text{ N/C})^2} = 133{,}600 \text{ N/C} = 1.34\times10^5 \text{ N/C}$$

and its angle below the $-x$-axis is $\theta = \tan^{-1}(|E_x/E_y|) = 44°$. We can also write $\vec{E}_{\text{net}} = (1.34\times10^5$ N/C, 136° CW from the $+x$-axis$)$.

27.31. Model: The electric field is that of three point charges $q_1 = -Q$, $q_2 = -Q$, and $q_3 = +4Q$.
Visualize: Assume the charges are in the x-y plane. The net electric field at point P is $\vec{E}_{\text{net}} = \vec{E}_1 + \vec{E}_2 + \vec{E}_3$. The procedure will be to find the magnitudes of the electric fields, to write them in component form, and to add the components.

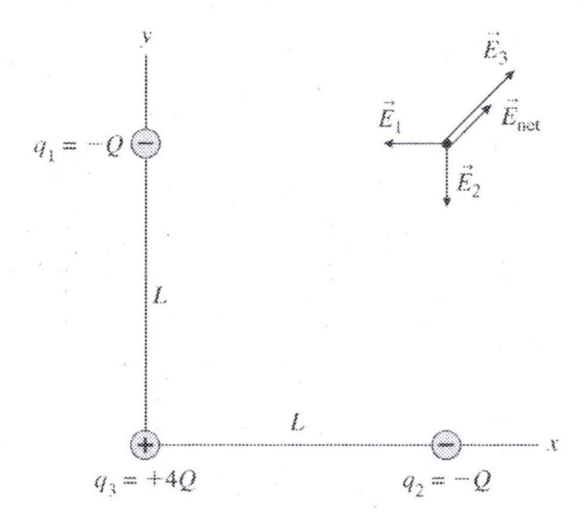

Solve: **(a)** The electric field produced by q_1 points toward q_1 and is given by

$$\vec{E}_1 = -\left(\frac{1}{4\pi\varepsilon_0}\frac{Q}{L^2}\right)\hat{i}$$

The electric field produced by q_2 points toward q_2 and is given by

$$\vec{E}_2 = -\left(\frac{1}{4\pi\varepsilon_0}\frac{Q}{L^2}\right)\hat{j}.$$

The electric field produced by q_3 is

$$E_3 = \frac{1}{4\pi\varepsilon_0}\left(\frac{4Q}{L^2+L^2}\right) = \frac{1}{4\pi\varepsilon_0}\left(\frac{2Q}{L^2}\right)$$

$\vec{E}_3$ points away from q_3 and makes an angle $\phi = \tan^{-1}(L/L) = 45°$ with the x-axis. So

$$\vec{E}_3 = \frac{1}{4\pi\varepsilon_0}\left(\frac{2Q}{L^2}\right)(\cos\phi\hat{i} + \sin\phi\hat{j}) = \frac{1}{4\pi\varepsilon_0}\frac{2Q}{L^2}\left(\frac{1}{\sqrt{2}}\hat{i} + \frac{1}{\sqrt{2}}\hat{j}\right)$$

Adding these three vectors gives

$$\vec{E}_{net} = \vec{E}_1 + \vec{E}_2 + \vec{E}_3 = \frac{1}{4\pi\varepsilon_0}\frac{Q}{L^2}\left[\left(\frac{2}{\sqrt{2}}-1\right)\hat{i} + \left(\frac{2}{\sqrt{2}}-1\right)\hat{j}\right] = \frac{1}{4\pi\varepsilon_0}\frac{Q}{L^2}(\hat{i}+\hat{j})(\sqrt{2}-1)$$

(b) The force on the charge is $F = ma = qE_{net}$. Therefore,

$$a = \frac{q}{m}\frac{1}{4\pi\varepsilon_0}\frac{Q(\sqrt{2}-1)\sqrt{2}}{L^2} = \frac{1}{4\pi\varepsilon_0}\left(\frac{0.586\,qQ}{mL^2}\right)$$

27.35. Model: The electric field is that of three point charges.
Visualize:

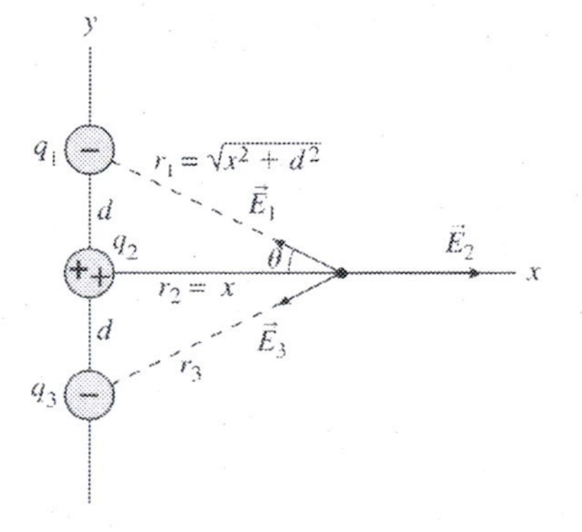

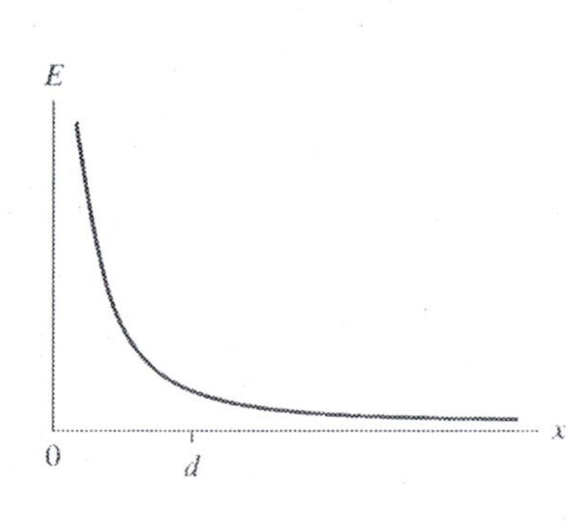

The field at points on the x-axis is $\vec{E}_{\text{net}} = \vec{E}_1 + \vec{E}_2 + \vec{E}_3$.

Solve: **(a)** We note from the symmetry that the y-component of $\vec{E}_1$ and $\vec{E}_3$ cancel. Since $\vec{E}_2$ has no y-component, the net field will have only an x-component. The x-components of $\vec{E}_1$ and $\vec{E}_3$ are equal, so

$$(E_{\text{net}})_x = E_2 - 2E_1\cos\theta \qquad (E_{\text{net}})_y = (E_{\text{net}})_z = 0\text{ N/C}$$

Note that the signs of q_1 and q_3 were used in writing this equation. The electric field strength due to q_2 is

$$E_2 = \frac{1}{4\pi\varepsilon_0}\frac{|q_2|}{r_2^2} = \frac{1}{4\pi\varepsilon_0}\frac{2q}{x^2}$$

The electric field strength due to q_1 is

$$E_1 = \frac{1}{4\pi\varepsilon_0}\frac{|q_1|}{r_1^2} = \frac{1}{4\pi\varepsilon_0}\frac{q}{x^2+d^2}$$

From geometry,

$$\cos\theta = \frac{x}{r_1} = \frac{x}{\sqrt{x^2+d^2}}$$

Assembling these pieces, the net field is

$$(E_{\text{net}})_x = \frac{1}{4\pi\varepsilon_0}\frac{2q}{x^2} - \frac{1}{4\pi\varepsilon_0}\frac{2q}{x^2+d^2}\frac{x}{\sqrt{x^2+d^2}} = \frac{2q}{4\pi\varepsilon_0}\left[\frac{1}{x^2} - \frac{x}{\left(x^2+d^2\right)^{3/2}}\right]$$

$$\Rightarrow \vec{E}_{\text{net}} = \frac{2q}{4\pi\varepsilon_0}\left[\frac{1}{x^2} - \frac{x}{\left(x^2+d^2\right)^{3/2}}\right]\hat{i}$$

(b) For $x << d$, the observation point is very close to $q_2 = +2q$. Furthermore, at $x \approx 0$ m the fields $\vec{E}_1$ and $\vec{E}_3$ are nearly opposite to each other and will nearly cancel. So for $x << d$ we expect the field to be that of a point charge $+2q$ at the origin. To test this prediction, we note that for $x << d$

$$\frac{x}{\left(x^2+d^2\right)^{3/2}} \approx \frac{x}{d^3} \approx 0 \Rightarrow \vec{E}_{\text{net}} = \frac{1}{4\pi\varepsilon_0}\frac{2q}{x^2}\hat{i}$$

This is, indeed, the field on the x-axis of point charge $2q$ at the origin. For $x >> d$, the three charges appear as a single charge of value $q_{\text{net}} = q_1 + q_2 + q_3 = 0$. So we expect $\vec{E} \approx 0$ when $x >> d$. In this limit,

$$\frac{1}{x^2} - \frac{x}{\left(x^2+d^2\right)^{3/2}} \approx \frac{1}{x^2} - \frac{x}{x^3} = \frac{1}{x^2} - \frac{1}{x^2} = 0$$

so the field does rapidly become zero, as expected.
(c) A graph of E_x is shown in the figure above.

27.37. Model: The electric field is that of two infinite lines of charge extending out of the page.
Visualize: Please refer to Figure P27.37. The line charges lie on the x-axis.
Solve: **(a)** From Equation 27.15, the electric field strength due to an infinite line of charge at a distance r from the line charge is

$$E = \frac{1}{4\pi\varepsilon_0}\frac{2|\lambda|}{r}$$

For the left and right line charges, the electric fields are

$$E_{\text{left}} = E_{\text{right}} = \frac{1}{4\pi\varepsilon_0}\frac{2\lambda}{\sqrt{y^2+(d/2)^2}} = \frac{1}{4\pi\varepsilon_0}\frac{4\lambda}{\sqrt{4y^2+d^2}}$$

$$\cos\phi = \frac{d/2}{\sqrt{y^2 + d^2/4}} = \frac{d}{\sqrt{4y^2 + d^2}} \qquad \sin\phi = \frac{2y}{\sqrt{4y^2 + d^2}}$$

$$\Rightarrow \vec{E}_{\text{left}} = \frac{1}{4\pi\varepsilon_0}\frac{4\lambda}{\sqrt{4y^2 + d^2}}\left(\frac{d}{\sqrt{4y^2 + d^2}}\hat{i} + \frac{2y}{\sqrt{4y^2 + d^2}}\hat{j}\right)$$

$$\vec{E}_{\text{right}} = \frac{1}{4\pi\varepsilon_0}\frac{4\lambda}{\sqrt{4y^2 + d^2}}\left(\frac{d}{\sqrt{4y^2 + d^2}}\hat{i} - \frac{2y}{\sqrt{4y^2 + d^2}}\hat{j}\right)$$

$$\Rightarrow \vec{E}_{\text{net}} = \vec{E}_{\text{left}} + \vec{E}_{\text{right}} = \frac{1}{4\pi\varepsilon_0}\left(\frac{4\lambda}{\sqrt{4y^2 + d^2}}\right)(2)\left(\frac{d}{\sqrt{4y^2 + d^2}}\right)\hat{i} = \frac{1}{4\pi\varepsilon_0}\frac{8\lambda d}{(4y^2 + d^2)}\hat{i}$$

Thus the field strength is

$$E = \frac{1}{4\pi\varepsilon_0}\frac{8\lambda d}{4y^2 + d^2}$$

(b) To draw a graph of E_{net} versus y, we calculated E_{net} at a few selected values of y.

y (in units of d)	$E_{\text{net}}\left(\text{in units of } \frac{1}{4\pi\varepsilon_0}\frac{\lambda}{d}\right)$
0	8
0.5	4
1.0	1.6
2.0	0.47
3.0	0.22
4.0	0.12

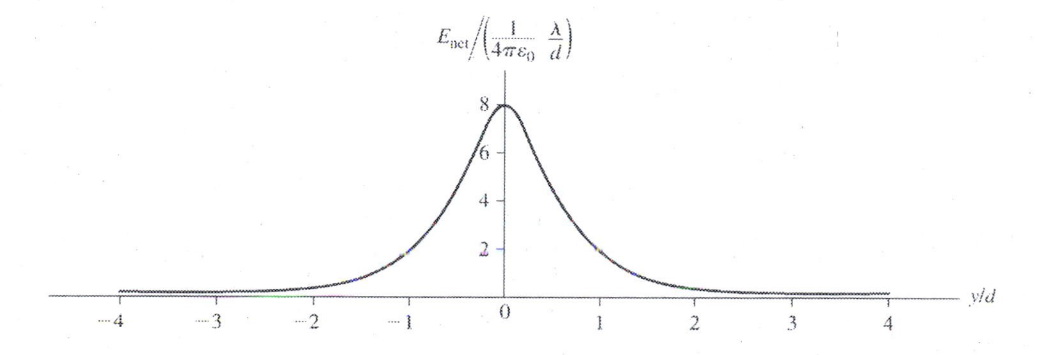

27.41. Model: The electric field is that of an infinite line charge.

Visualize:

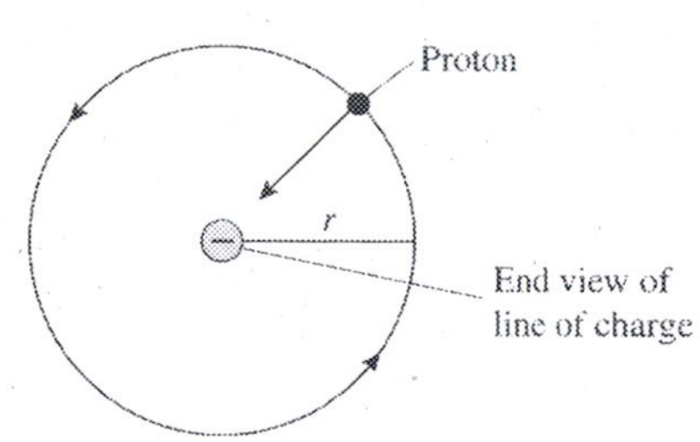

Solve: The wire must be negative to attract the proton. From Equation 27.15, the electric field strength due to an infinite line of charge at a distance r from the line charge is

$$E = \frac{1}{4\pi\varepsilon_0}\frac{2|\lambda|}{r}$$

The force on the proton due to this electric field causes the centripetal acceleration:

$$F = qE = \frac{1}{4\pi\epsilon_0}\frac{2q|\lambda|}{r} = \frac{mv^2}{r} \Rightarrow |\lambda| = (4\pi\epsilon_0)\frac{mv^2}{2q}$$

Using $v = 2\pi r/T = 2\pi rf$,

$$|\lambda| = (4\pi\epsilon_0)\frac{m(4\pi^2r^2f^2)}{2q} = \frac{(1.67\times10^{-27}\text{ kg})4\pi^2(1.0\times10^{-2}\text{ m})^2(1.0\times10^6\text{ s}^{-1})^2}{(9.0\times10^9\text{ N m}^2/\text{C}^2)2(1.60\times10^{-19}\text{ C})} = 2.3\times10^{-9}\text{ C/m}$$

Because the wire has to be negative, $\lambda = -2.3$ nC/m.

27.43. Model: The electric field is that of a line charge of length L.
Visualize: Please refer to Figure P27.43. Let the bottom end of the rod be the origin of the coordinate system. Divide the rod into many small segments of charge Δq and length $\Delta y'$. Segment i creates a small electric field at the point P that makes an angle θ with the horizontal. The field has both x and y components, but $E_z = 0$ N/C. The distance to segment i from point P is $(x^2 + y'^2)^{1/2}$.
Solve: The electric field created by segment i at point P is

$$\vec{E}_i = \frac{\Delta q}{4\pi\epsilon_0(x^2+y'^2)}(\cos\theta\hat{i} - \sin\theta\,\hat{j}) = \frac{\Delta q}{4\pi\epsilon_0(x^2+y'^2)}\left(\frac{x}{\sqrt{x^2+y'^2}}\hat{i} - \frac{y'}{\sqrt{x^2+y'^2}}\hat{j}\right)$$

The net field is the sum of all the $\vec{E}_i$, which gives $\vec{E} = \sum_i \vec{E}_i$. Δq is not a coordinate, so before converting the sum to an integral we must relate charge Δq to length $\Delta y'$. This is done through the linear charge density $\lambda = Q/L$, from which we have the relationship

$$\Delta q = \lambda\Delta y' = \frac{Q}{L}\Delta y'$$

With this charge, the sum becomes

$$\vec{E} = \frac{Q/L}{4\pi\epsilon_0}\sum_i\left[\frac{x\Delta y'}{(x^2+y'^2)^{3/2}}\hat{i} - \frac{y'\Delta y'}{(x^2+y'^2)^{3/2}}\hat{j}\right]$$

Now we let $\Delta y' \to dy'$ and replace the sum by an integral from $y' = 0$ m to $y' = L$. Thus,

$$\vec{E} = \frac{(Q/L)}{4\pi\epsilon_0}\left(\int_0^L\frac{x\,dy'}{(x^2+y'^2)^{3/2}}\hat{i} - \int_0^L\frac{y'dy'}{(x^2+y'^2)^{3/2}}\hat{j}\right) = \frac{(Q/L)}{4\pi\epsilon_0}\left(x\left[\frac{y'}{x^2\sqrt{x^2+y'^2}}\right]_0^L\hat{i} - \left[\frac{-1}{\sqrt{x^2+y'^2}}\right]_0^L\hat{j}\right)$$

$$= \frac{Q}{4\pi\epsilon_0}\frac{1}{x\sqrt{x^2+L^2}}\hat{i} - \frac{1}{4\pi\epsilon_0}\left(\frac{Q}{Lx}\right)\left(1 - \frac{x}{\sqrt{x^2+L^2}}\right)\hat{j}$$

27.47. Model: Assume that the quarter-circle plastic rod is thin and that the charge lies along the quarter-circle of radius R.
Visualize:

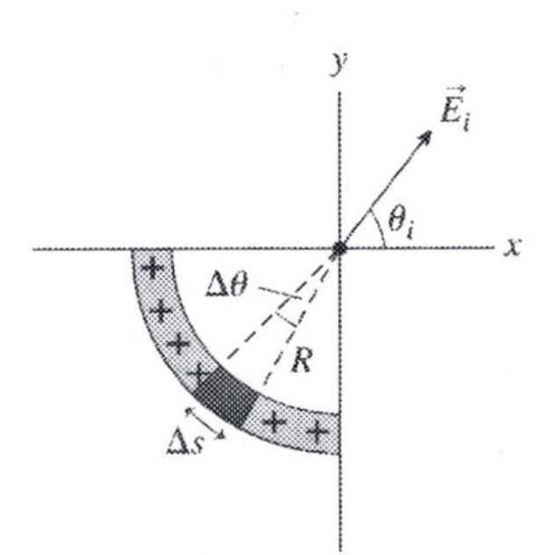

The origin of the coordinate system is at the center of the circle. Divide the rod into many small segments of charge Δq and arc length Δs.

Solve: **(a)** Segment i creates a small electric field $\vec{E}_i$ at the origin with two components:

$$(E_i)_x = E_i \cos\theta_i \qquad (E_i)_y = E_i \sin\theta_i$$

Note that the angle θ_i depends on the location of the segment i. Now all segments are at distance $r_i = R$ from the origin, so

$$E_i = \frac{1}{4\pi\varepsilon_0}\frac{\Delta q}{r_i^2} = \frac{1}{4\pi\varepsilon_0}\frac{\Delta q}{R^2}$$

The linear charge density of the rod is $\lambda = Q/L$, where L is the rod's length (L = quarter-circumference = $\pi R/2$). This allows us to relate charge Δq to the arc length Δs through

$$\Delta q = \lambda\Delta s = \left(\frac{Q}{L}\right)\Delta s = \left(\frac{2Q}{\pi R}\right)\Delta s$$

Using $\Delta s = R\Delta\theta$, the components of the electric field at the origin are

$$(E_i)_x = \frac{1}{4\pi\varepsilon_0}\frac{\Delta q}{R^2}\cos\theta_i = \frac{1}{4\pi_0}\frac{1}{R^2}\left(\frac{2Q}{\pi R}\right)R\Delta\theta\cos\theta_i = \frac{1}{4\pi\varepsilon_0}\left(\frac{2Q}{\pi R^2}\right)\Delta\theta\cos\theta_i$$

$$(E_i)_y = \frac{1}{4\pi\varepsilon_0}\frac{\Delta q}{R^2}\sin\theta_i = \frac{1}{4\pi\varepsilon_0}\frac{1}{R^2}\left(\frac{2Q}{\pi R}\right)R\Delta\theta\sin\theta_i = \frac{1}{4\pi\varepsilon_0}\left(\frac{2Q}{\pi R^2}\right)\Delta\theta\sin\theta_i$$

(b) The x- and y-components of the electric field for the entire rod are the integrals of the expressions in part (a) from $\theta = 0$ rad to $\theta = \pi/2$. We have

$$E_x = \frac{1}{4\pi\varepsilon_0}\left(\frac{2Q}{\pi R^2}\right)\int_0^{\pi/2}\cos\theta\, d\theta \qquad E_y = \frac{1}{4\pi\varepsilon_0}\left(\frac{2Q}{\pi R^2}\right)\int_0^{\pi/2}\sin\theta\, d\theta$$

(c) The integrals are

$$\int_0^{\pi/2}\sin\theta d\theta = [-\cos\theta]_0^{\pi/2} = -\left(\cos\frac{\pi}{2} - \cos 0\right) = +1 \qquad \int_0^{\pi/2}\cos\theta d\theta = [\sin\theta]_0^{\pi/2} = \sin\frac{\pi}{2} - \sin 0 = +1$$

The electric field is

$$\vec{E} = \frac{1}{4\pi\varepsilon_0}\frac{2Q}{\pi R^2}(\hat{i} + \hat{j})$$

27.51. **Model:** The parallel plates form a parallel-plate capacitor. The electric field inside a parallel-plate capacitor is a uniform field, so the electron and proton will have constant acceleration.
Visualize:

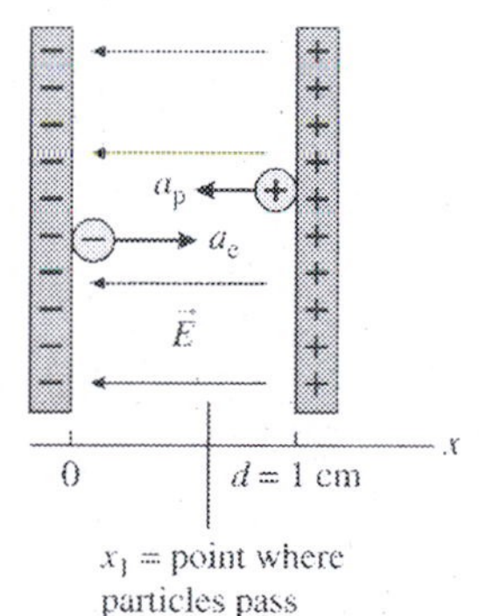

The negative plate is at $x = 0$ m and the positive plate is at $x = d = 1$ cm.
Solve: Both particles accelerate from rest ($v_0 = 0$ m/s), so at time t their positions are

$$x_e = x_{e0} + \frac{1}{2}a_e t^2 = \frac{1}{2}a_e t^2 \qquad x_p = x_{p0} + \frac{1}{2}a_p t^2 = d + \frac{1}{2}a_p t^2$$

At some instant of time t_1 the electron and proton have the same position: $x_e = x_p = x_1$. This is the point where they pass each other. At this instant,

$$x_1 = \frac{1}{2}a_e t_1^2 \qquad x_1 = d + \frac{1}{2}a_p t_1^2$$

These are two equations in the two unknowns x_1 and t_1. From the first equation, $\frac{1}{2}t_1^2 = x_1/a_e$. Using this in the second equation gives

$$x_1 = d + \frac{a_p}{a_e}x_1 \Rightarrow x_1 = \frac{d}{1 + a_p/a_e}$$

To finish, we need to find the accelerations of the electron and proton. Both particles are in a parallel-plate capacitor with $E_{cap} = Q/\varepsilon_0 A$. The field points to the left, so $E_x = -Q/\varepsilon_0 A$. The proton's acceleration is

$$a_p = \frac{F_p}{m_p} = \frac{q_p E_x}{m_p} = \frac{e(-Q/\varepsilon_0 A)}{m_p} = -\frac{eQ/\varepsilon_0 A}{m_p}$$

The proton's acceleration is negative, as expected. For the electron,

$$a_e = \frac{F_e}{m_e} = \frac{q_e E_x}{m_e} = \frac{-e(-Q/\varepsilon_0 A)}{m_e} = \frac{eQ/\varepsilon_0 A}{m_e}$$

Consequently, the acceleration ratio is $a_p/a_e = m_e/m_p$. Using this, the point where the two charges pass is

$$x_1 = \frac{d}{1 + m_e/m_p} = \frac{1 \text{ cm}}{1 + 9.11\times10^{-31} \text{ kg}/1.67\times10^{-27} \text{ kg}} = 0.9995 \text{ cm}$$

Assess: This is very close to where the proton starts. Since there is a factor of ~1800 difference in the masses of the proton and electron, the electron accelerates much faster than the proton.

27.55. Model: Assume that the plates form a parallel-plate capacitor. The electric field inside the capacitor is uniform, so that the protons will have a constant acceleration.
Visualize:

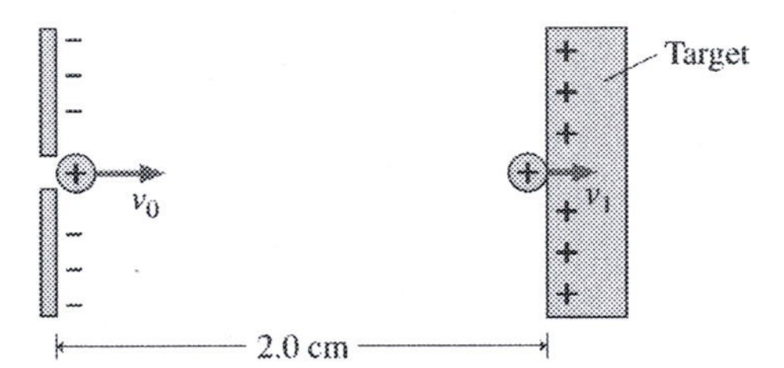

Solve: **(a)** The acceleration needed to slow the protons to 2.0×10^5 m/s can be obtained using the kinematic equation of motion $v_1^2 = v_0^2 + 2a(x_1 - x_0)$. We have

$$a = \frac{v_1^2 - v_0^2}{2(x_1 - x_0)} = \frac{(2.0\times10^5 \text{ m/s})^2 - (2.0\times10^6 \text{ m/s})^2}{2(0.020 \text{ m})} = -9.9\times10^{13} \text{ m/s}^2$$

The force on the proton due to the electric field between the plates is $F = ma = qE_x$. The electric field is

$$E_x = \frac{ma}{q} = \frac{(1.67\times10^{-27} \text{ kg})(-9.9\times10^{13} \text{ m/s}^2)}{+1.6\times10^{-19} \text{ C}} = -1.033\times10^6 \text{ N/C}$$

The minus sign indicates that the field points to the left. Because $E = \eta/\varepsilon_0$ for a parallel-plate capacitor, where E is the field strength,

$$\eta = \varepsilon_0 E = (8.85\times10^{-12} \text{ C}^2/\text{Nm}^2)(1.033\times10^6 \text{ N/C}) = 9.1\times10^{-6} \text{ C/m}^2$$

It is clear that the electric field must oppose the motion of the protons. This requires that the first plate be negatively charged and the second plate be positively charged.
(b) Let us calculate E that makes protons come to rest completely. Combining the equations for E and a in part (a),

$$E=\frac{m(v_1^2-v_0^2)}{2q(x_1-x_0)}=\frac{(1.67\times10^{-27}\text{ kg})\left[(0\text{ m/s})^2-(2\times10^6\text{ m/s})^2\right]}{2(1.6\times10^{-19}\text{ C})(0.020\text{ m})}=1.04\times10^6\text{ N/C}$$

$$\Rightarrow\eta=\varepsilon_0E=(8.85\times10^{-12}\text{ C}^2/\text{Nm}^2)(1.04\times10^6\text{ N/C})=9.24\times10^{-6}\text{ C/m}^2$$

Because the surface charge density of $\pm1.0\times10^{-5}$ C/m^2 in your device is larger than 9.24×10^{-6} C/m^2, the protons will not reach the positive plate. The beam will in fact reverse its direction. The protons will not be stopped, so no, the device does not work.

27.59. Model: The electron orbiting the proton experiences a force given by Coulomb's law.
Visualize:

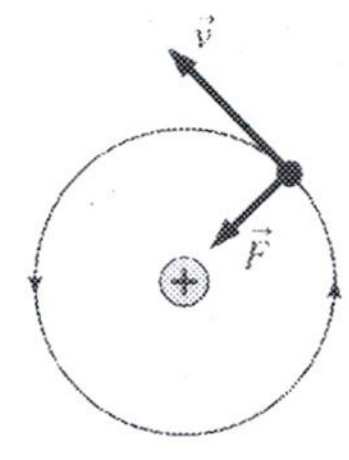

Solve: The force that causes the circular motion is

$$F=\frac{1}{4\pi\varepsilon_0}\frac{q_p|q_e|}{r^2}=\frac{m_ev^2}{r}=\frac{m_e(4\pi^2r^2f^2)}{r}$$

where we used $v=2\pi r/T=2\pi rf$. The radius is

$$r=\left[\frac{1}{4\pi\varepsilon_0}\frac{q_p|q_e|}{m_e(4\pi^2f^2)}\right]^{1/3}$$

$$=\left[(9.0\times10^9\text{ N m}^2/\text{C}^2)\frac{(1.60\times10^{-19}\text{ C})(1.60\times10^{-19}\text{ C})}{(9.1\times10^{-31}\text{ kg})4\pi^2(1.0\times10^{12}\text{ s}^{-1})^2}\right]^{1/3}=1.86\times10^{-8}\text{ m}=18.6\text{ nm}$$

27.63. Solve: (a) A very long charged wire has linear charge density 2.0×10^{-7} C/m. At what distance from the wire is the field strength 25,000 N/C?
(b) Solving for r,

$$r=\frac{(9\times10^9\text{ N m}^2/\text{C}^2)2(2.0\times10^{-7}\text{ C/m})}{25{,}000\text{ N/C}}=0.144\text{ m}=14.4\text{ cm}$$

27.65. Solve: (a) A proton is released from the positive plate of a parallel-plate capacitor and accelerates toward the negative plate at 2.0×10^{12} m/s^2. If the capacitor plates are 2.0 cm × 2.0 cm squares, what is the magnitude of the charge on each?
(b) Solve the first equation for E and substitute into the second equation. The charge is

$$Q=\frac{(1.67\times10^{-27}\text{ kg})(2.0\times10^{12}\text{ m/s}^2)(8.85\times10^{-12}\text{ C}^2/\text{N m}^2)(0.020\text{ m})^2}{1.60\times10^{-19}\text{ C}}=7.4\times10^{-11}\text{ C}$$

28

GAUSS'S LAW

Exercises and Problems

28.1. Visualize:

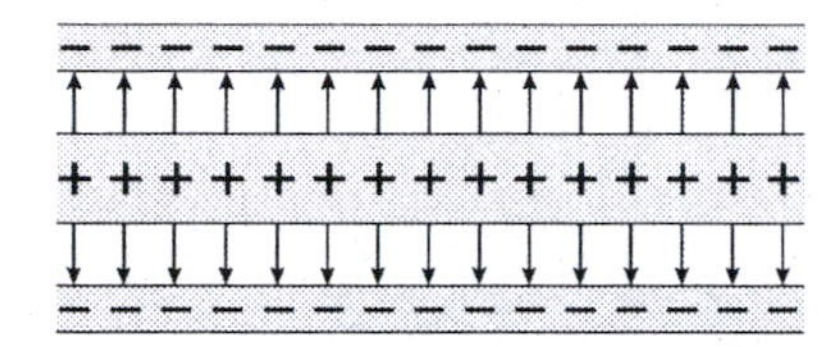

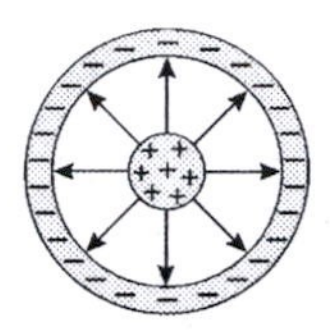

As discussed in Section 28.1, the symmetry of the electric field must match the symmetry of the charge distribution. In particular, the electric field of a cylindrically symmetric charge distribution cannot have a component parallel to the cylinder axis. Also, the electric field of a cylindrically symmetric charge distribution cannot have a component tangent to the circular cross section. The only shape for the electric field that matches the symmetry of the charge distribution with respect to (i) translation parallel to the cylinder axis, (ii) rotation by an angle about the cylinder axis, and (iii) reflections in any plane containing or perpendicular to the cylinder axis is the one shown in the figure.

28.7. Model: The electric flux "flows" *out* of a closed surface around a region of space containing a net positive charge and *into* a closed surface surrounding a net negative charge.
Visualize: Please refer to Figure EX28.7. Let A be the area of each of the six faces of the cube.
Solve: The electric flux is defined as $\Phi_e = \vec{E}\cdot\vec{A} = EA\cos\theta$, where θ is the angle between the electric field and a line *perpendicular* to the plane of the surface. The electric flux out of the closed cube surface is

$$\Phi_{out} = (20\text{ N/C} + 10\text{ N/C})A\cos 0° = (30A)\text{ N m}^2\text{/C}$$

Similarly, the electric flux into the closed cube surface is

$$\Phi_{in} = (20\text{ N/C} + 15\text{ N/C} + 10\text{ N/C})A\cos 180° = -(45A)\text{ N m}^2\text{/C}$$

Because the cube contains positive charge, $\Phi_{out} + \Phi_{in}$ must be positive. This means $\Phi_{out} + \Phi_{in} + \Phi_{unknown} > 0\text{ N m}^2\text{/C}$. Therefore,

$$(30A)\text{ N m}^2\text{/C} + (-45A)\text{ N m}^2\text{/C} + \Phi_{unknown} > 0\text{ N m}^2\text{/C}$$

$$\Rightarrow \Phi_{unknown} > (15A)\text{ N m}^2\text{/C}$$

That is, the unknown vector points *out of* the front face of the cube and its field strength is greater than 15 N/C.

28.13. Model: The electric field over the rectangle in the xz plane is uniform.
Solve: **(a)** The area vector is perpendicular to the xz plane. Thus

$$\vec{A} = (2.0\text{ cm} \times 3.0\text{ cm})\hat{j} = (6.0\times10^{-4}\text{ m}^2)\hat{j}$$

The electric flux through the rectangle is

$$\Phi_e = \vec{E}\cdot\vec{A} = (50\hat{i} + 100\hat{k})\text{ N/C}\cdot(6.0\times10^{-4}\text{ m}^2)\hat{j}$$

$$= (300\times10^{-4}\text{ N m}^2\text{/C})(\hat{i}\cdot\hat{j}) + (600\times10^{-4}\text{ N m}^2\text{/C})(\hat{k}\cdot\hat{j}) = 0\text{ N m}^2\text{/C}$$

(b) The flux is

$$\begin{aligned}\Phi_e &= \vec{E}\cdot\vec{A} = \left(50\hat{\imath}+100\hat{\jmath}\right)\text{ N/C}\cdot\left(6.0\times10^{-4}\text{ m}^2\right)\hat{\jmath}\\ &= \left(300\times10^{-4}\text{ N m}^2\text{/C}\right)\left(\hat{\imath}\cdot\hat{\jmath}\right)+\left(600\times10^{-4}\text{ N m}^2\text{/C}\right)\hat{\jmath}\cdot\hat{\jmath}\\ &= 0\text{ N m}^2\text{/C}+\left(6.0\times10^{-2}\text{ N m}^2\text{/C}\right) = 6.0\times10^{-2}\text{ N m}^2\text{/C}\end{aligned}$$

28.15. Model: The electric field over the two faces of the box perpendicular to $\vec{E}$ is uniform.
Visualize:

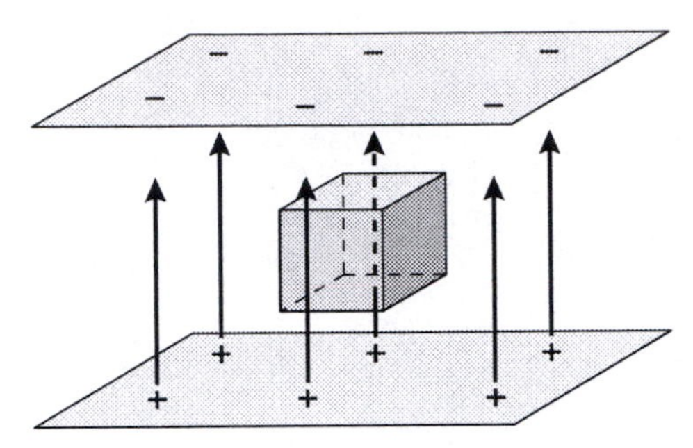

Solve: The electric field is parallel to four sides of the box, so the electric fluxes through these four surfaces are zero. The field is out of the top surface, so $\Phi_{top} = EA$. The field is into the bottom surface, so $\Phi_{bottom} = -EA$. Thus the net flux is $\Phi_e = 0\text{ N m}^2\text{/C}^2$.

28.17. Visualize:

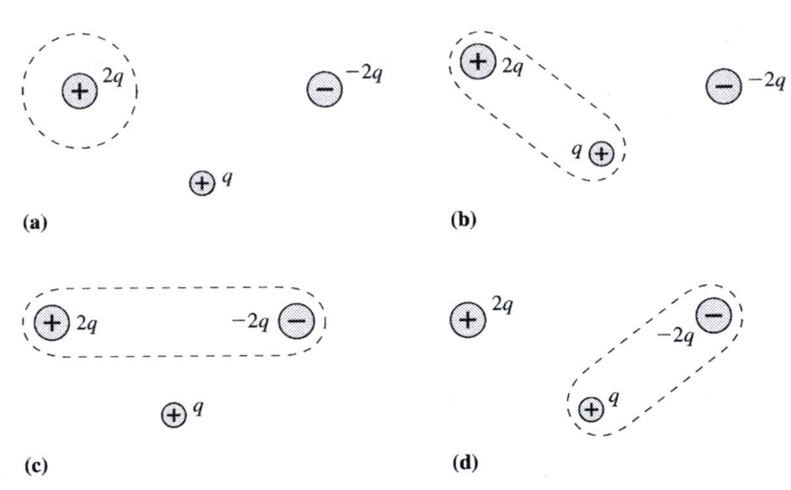

For *any* closed surface that encloses a total charge Q_{in}, the net electric flux through the closed surface is $\Phi_e = Q_{in}/\varepsilon_0$.

28.25. Model: The excess charge on a conductor resides on the outer surface.
Solve: The electric field at the surface of a charged conductor is

$$\vec{E}_{surface} = \left(\frac{\eta}{\varepsilon_0},\text{ perpendicular to surface}\right)$$

$$\Rightarrow \eta = \varepsilon_0 E_{surface} = \left(8.85\times10^{-12}\text{ C}^2\text{/Nm}^2\right)\left(3.0\times10^6\text{ N/C}\right) = 2.7\times10^{-5}\text{ C/m}^2$$

Assess: It is the air molecules just above the surface that "break down" when the E-field becomes strong enough to accelerate stray charges to approximately 15 eV between collisions, thus causing collisional ionization. It does not make any difference whether E points toward or away from the surface.

28.31. Model: Because the tetrahedron contains no charge, the net flux through the tetrahedron is zero.
Visualize:

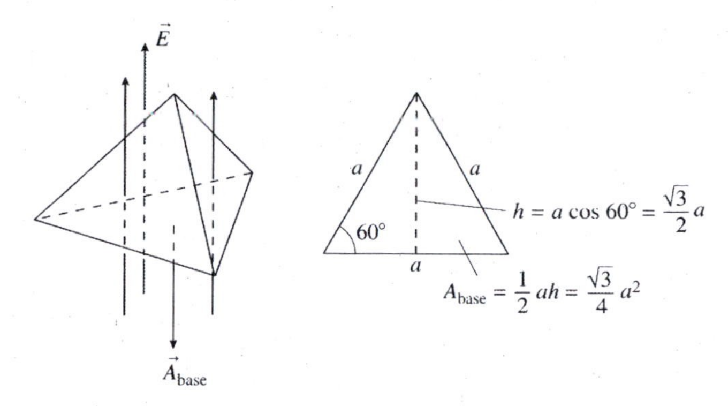

Solve: **(a)** The area of the base of the tetrahedron is $\frac{\sqrt{3}}{4}a^2$, where $a = 20$ cm is the length of one of the sides. Because the base of the tetrahedron is parallel to the ground and the vertical uniform electric field passes upward through the base, the angle between $\vec{E}$ and $\vec{A}$ is 180°. Thus,

$$\Phi_{base} = \vec{E}\cdot\vec{A}_{base} = EA_{base}\cos 180° = -E\frac{\sqrt{3}}{4}a^2 = -(200\text{ N/C})\frac{\sqrt{3}}{4}(0.20\text{ m})^2 = -3.46\text{ N m}^2/\text{C}$$

The electric flux through the base is -3.5 N m^2/C.

(b) Since $\vec{E}$ is perpendicular to the base, the other three sides of the tetrahedron share the flux equally. Because the tetrahedron contains no charge,

$$\Phi_{net} = \Phi_{base} + 3\Phi_{side} = 0\text{ N m}^2/\text{C} \Rightarrow -3.46\text{ N m}^2/\text{C} + 3\Phi_{side} = 0\text{ N m}^2/\text{C}$$

$$\Rightarrow \Phi_{side} = 1.15\text{ N m}^2/\text{C}$$

28.35. Solve: **(a)** The electric field is

$$\vec{E} = (5000\,r^2)\hat{r}\text{ N/C} = 5000\,(0.20)^2\,\hat{r}\text{ N/C} = 200\hat{r}\text{ N/C}$$

So the electric field strength is 200 N/C.

(b) The area of the surface is $A_{sphere} = 4\pi(0.20\text{ m})^2 = 0.5027\text{ m}^2$. Thus, the electric flux is

$$\Phi_e = \oint\vec{E}\cdot d\vec{A} = EA_{sphere} = (200\text{ N/C})(0.5027\text{ m}^2) = 100.5\text{ N m}^2/\text{C} \approx 101\text{ N m}^2/\text{C}$$

(c) Because $\Phi_e = Q_{in}/\varepsilon_0$,

$$Q_{in} = \varepsilon_0\Phi_e = (8.85\times10^{-12}\text{ C}^2/\text{N m}^2)(100.5\text{ N m}^2/\text{C}) = 8.9\times10^{-10}\text{ C}$$

28.37. Model: The excess charge on a conductor resides on the outer surface.
Visualize:

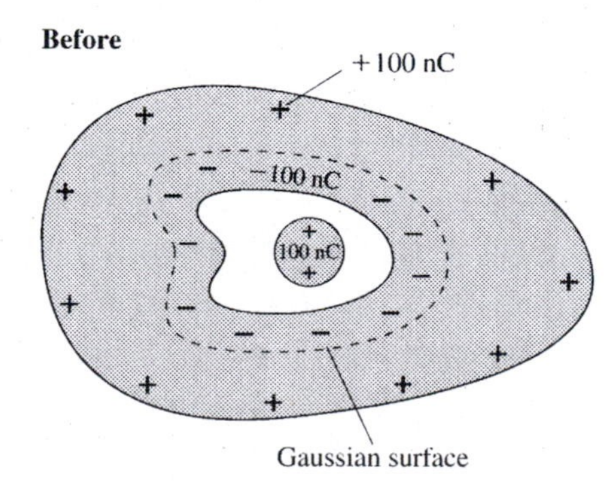

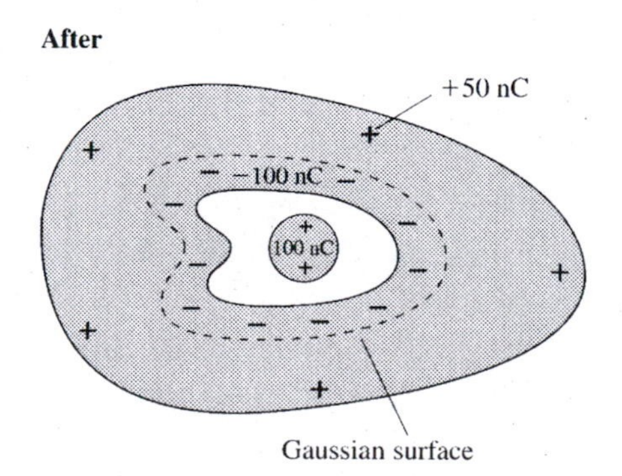

Solve: **(a)** Consider a Gaussian surface surrounding the cavity just *inside* the conductor. The electric field inside a conductor in electrostatic equilibrium is zero, so $\vec{E}$ is zero at all points on the Gaussian surface. Thus $\Phi_e = 0$. Gauss's law tells us that $\Phi_e = Q_{in}/\varepsilon_0$, so the net charge enclosed by this Gaussian surface is $Q_{in} = Q_{point} + Q_{wall} = 0$. We know that $Q_{point} = +100$ nC, so $Q_{wall} = -100$ nC. The positive charge in the cavity attracts an equal negative charge to the inside surface.
(b) The conductor started out neutral. If there is −100 nC on the wall of the cavity, then the exterior surface of the conductor was initially +100 nC. Transferring −50 nC to the conductor reduces the exterior surface charge by 50 nC, leaving it at +50 nC.
Assess: The electric field inside the conductor stays zero.

28.41. Model: The charge distribution at the surface of the earth is assumed to be uniform and to have spherical symmetry.
Visualize:

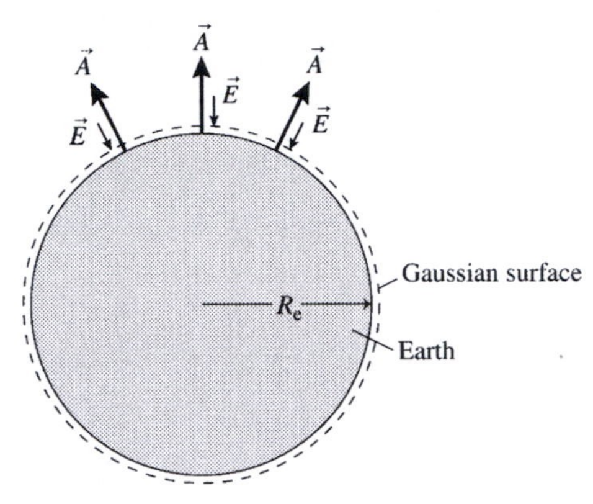

Due to the symmetry of the charge distribution, $\vec{E}$ is perpendicular to the Gaussian surface and the field strength has the same value at all points on the surface.
Solve: Gauss's law is $\Phi_e = \oint \vec{E} \cdot d\vec{A} = Q_{in}/\varepsilon_0$. The electric field points inward (negative flux), hence

$$Q_{in} = -\varepsilon_0 E A_{sphere} = -\left(8.85\times10^{-12}\ \text{C}^2/\text{N m}^2\right)\left(100\ \text{N/C}\right)4\pi\left(6.37\times10^6\ \text{m}\right)^2 = -4.51\times10^5\ \text{C}$$

28.45. Model: The hollow plastic ball has a charge uniformly distributed on its outer surface. This distribution leads to a spherically symmetric electric field.
Visualize:

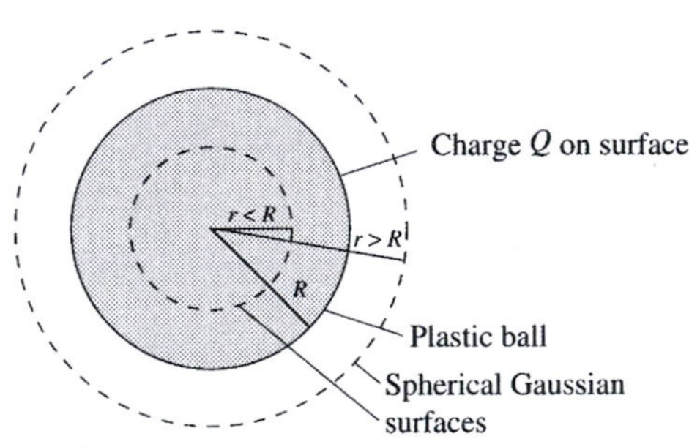

The figure shows Gaussian surfaces at $r < R$ and $r > R$.
Solve: **(a)** Gauss's law for the Gaussian surface for $r < R$ where $Q_{in} = 0$ is

$$\Phi_e = \oint \vec{E} \cdot d\vec{A} = \frac{Q_{in}}{\varepsilon_0} = 0\ \text{N m}^2/\text{C} \Rightarrow E = 0\ \text{N/C}$$

(b) Gauss's law for the Gaussian surface for $r > R$ is

$$\Phi_e = \oint \vec{E} \cdot d\vec{A} = EA_{sphere} = \frac{Q_{in}}{\varepsilon_0} = \frac{Q}{\varepsilon_0} \Rightarrow EA_{sphere} = \frac{Q}{\varepsilon_0} \Rightarrow E = \frac{Q}{\varepsilon_0 A_{sphere}} = \frac{1}{4\pi\varepsilon_0}\frac{Q}{r^2}$$

Assess: A uniform spherical shell of charge has the same electric field at $r > R$ as a point charge placed at the center of the sphere. Additionally, the shell of charge exerts no electric force on a charged particle inside the shell.

28.49. Model: The infinitely wide plane of charge with surface charge density η polarizes the infinitely wide conductor.

Visualize:

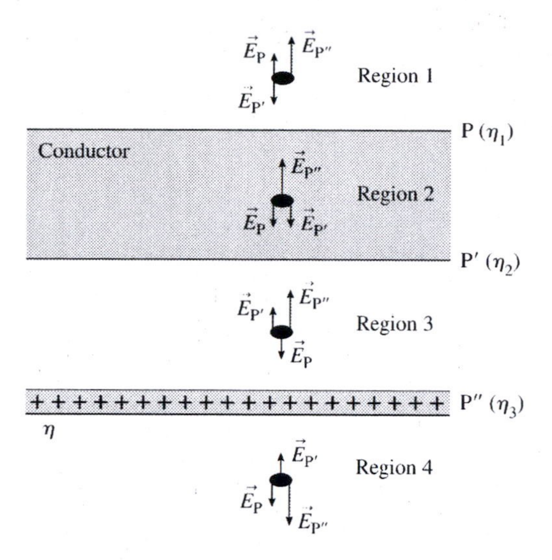

Because $\vec{E} = \vec{0}$ in the metal there will be an induced charge polarization. The face of the conductor adjacent to the plane of charge is negatively charged. This makes the other face of the conductor positively charged. We thus have three infinite planes of charge. These are P (top conducting face), P′ (bottom conducting face), and P″(plane of charge).

Solve: Let η_1, η_2, and η_3 be the surface charge densities of the three surfaces with η_2 a negative number. The electric field due to a plane of charge with surface charge density η is $E = \eta/2\varepsilon_0$. Because the electric field inside a conductor is zero (region 2),

$$\vec{E}_{\text{P}} + \vec{E}_{\text{P}'} + \vec{E}_{\text{P}''} = \vec{0}\ \text{N/C} \Rightarrow -\frac{\eta_1}{2\varepsilon_0}\hat{j} + \frac{\eta_2}{2\varepsilon_0}\hat{j} + \frac{\eta_3}{2\varepsilon_0}\hat{j} = \vec{0}\ \text{N/C} \Rightarrow -\eta_1 + \eta_2 + \eta = 0\ \text{C/m}^2$$

We have made the substitution $\eta_3 = \eta$. Also note that the field inside the conductor is downward from planes P and P′ and upward from P″. Because $\eta_1 + \eta_2 = 0\ \text{C/m}^2$, because the conductor is neutral, $\eta_2 = -\eta_1$. The above equation becomes

$$-\eta_1 - \eta_1 + \eta = 0\ \text{C/m}^2 \Rightarrow \eta_1 = \tfrac{1}{2}\eta \Rightarrow \eta_2 = -\tfrac{1}{2}\eta$$

We are now in a position to find electric field in regions 1–4.

For region 1,

$$\vec{E}_{\text{P}} = \frac{\eta}{4\varepsilon_0}\hat{j} \qquad \vec{E}_{\text{P}'} = -\frac{\eta}{4\varepsilon_0}\hat{j} \qquad \vec{E}_{\text{P}''} = \frac{\eta}{2\varepsilon_0}\hat{j}$$

The electric field is $\vec{E}_{\text{net}} = \vec{E}_{\text{P}} + \vec{E}_{\text{P}'} + \vec{E}_{\text{P}''} = (\eta/2\varepsilon_0)\hat{j}$.

In region 2, $\vec{E}_{\text{net}} = \vec{0}$ N/C. In region 3,

$$\vec{E}_{\text{P}} = -\frac{\eta}{4\varepsilon_0}\hat{j} \qquad \vec{E}_{\text{P}'} = \frac{\eta}{4\varepsilon_0}\hat{j} \qquad \vec{E}_{\text{P}''} = \frac{\eta}{2\varepsilon_0}\hat{j}$$

The electric field is $\vec{E}_{\text{net}} = (\eta/2\varepsilon_0)\hat{j}$.

In region 4,

$$\vec{E}_{\text{P}} = -\frac{\eta}{4\varepsilon_0}\hat{j} \qquad \vec{E}_{\text{P}'} = \frac{\eta}{4\varepsilon_0}\hat{j} \qquad \vec{E}_{\text{P}''} = -\frac{\eta}{2\varepsilon_0}\hat{j}$$

The electric field is $\vec{E}_{\text{net}} = -(\eta/2\varepsilon_0)\hat{j}$.

28.53. Model: The charge distribution in the shell has spherical symmetry.

Visualize:

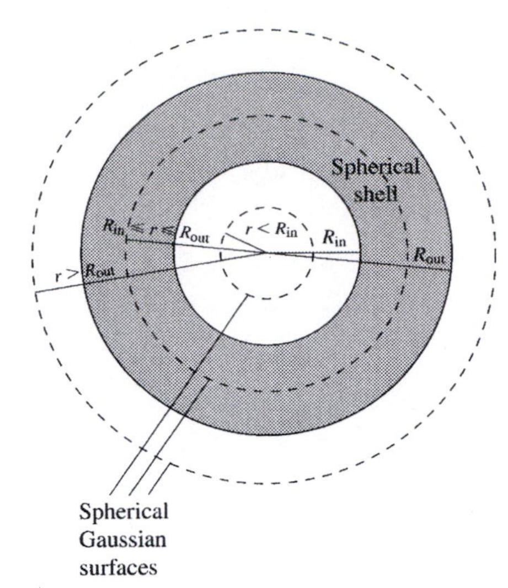

The spherical surfaces of radii $r \geq R_{out}$, $r \leq R_{in}$, and $R_{in} \leq r \leq R_{out}$, concentric with the spherical shell, are Gaussian surfaces.

Solve: **(a)** Gauss's law for the Gaussian surface $r \geq R_{out}$ is

$$\oint \vec{E}\cdot d\vec{A} = \frac{Q_{in}}{\varepsilon_0} \Rightarrow E\left(4\pi r^2\right) = \frac{Q}{\varepsilon_0} \Rightarrow E = \frac{1}{4\pi\varepsilon_0}\frac{Q}{r^2} \Rightarrow \vec{E} = \frac{1}{4\pi\varepsilon_0}\frac{Q}{r^2}\hat{r}$$

The vector form comes from the fact that the field is directed radially outward.

(b) For $r \leq R_{in}$, Gauss's law is

$$\oint \vec{E}\cdot d\vec{A} = \frac{Q_{in}}{\varepsilon_0} = \frac{0\text{ C}}{\varepsilon_0} \Rightarrow \vec{E} = \vec{0}$$

(c) For $R_{in} \leq r \leq R_{out}$, Gauss's law is

$$\oint \vec{E}\cdot d\vec{A} = \frac{Q_{in}}{\varepsilon_0} \Rightarrow E\left(4\pi r^2\right) = \frac{Q_{in}}{\varepsilon_0}$$

$$Q_{in} = \frac{4\pi}{3}\left(r^3 - R_{in}^3\right)\rho = \frac{4\pi}{3}\frac{\left(r^3 - R_{in}^3\right)Q}{\frac{4\pi}{3}\left(R_{out}^3 - R_{in}^3\right)} = Q\left(\frac{r^3 - R_{in}^3}{R_{out}^3 - R_{in}^3}\right)$$

$$\Rightarrow E = \frac{1}{4\pi r^2}\frac{Q}{\varepsilon_0}\frac{r^3 - R_{in}^3}{R_{out}^3 - R_{in}^3} \Rightarrow \vec{E} = \frac{1}{4\pi\varepsilon_0}\frac{Q}{r^2}\left(\frac{r^3 - R_{in}^3}{R_{out}^3 - R_{in}^3}\right)\hat{r}$$

(d) The result obtained in part (c) for the electric field simplifies to $\vec{E} = \vec{0}$, when $r = R_{in}$ which is the result obtained in part (b). Furthermore, at $r = R_{out}$, the electric field obtained in part (c) becomes

$$\vec{E} = \frac{1}{4\pi\varepsilon_0}\frac{Q}{R_{out}^2}\hat{r}$$

which is the same as the electric field obtained in part (a).

(e)

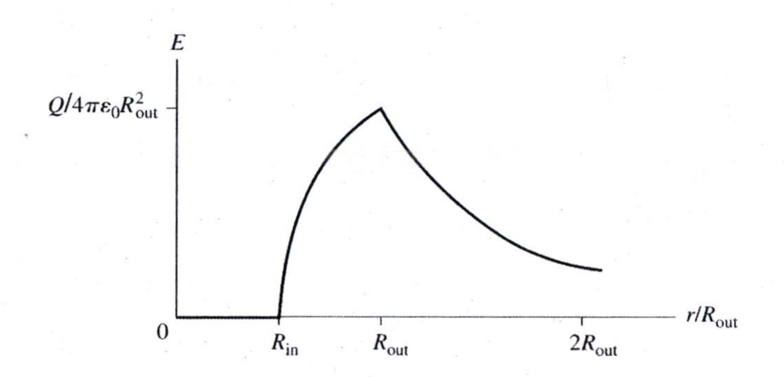

29

THE ELECTRIC POTENTIAL

Exercises and Problems

29.3. Model: The mechanical energy of the proton is conserved. A parallel-plate capacitor has a uniform electric field.
Visualize:

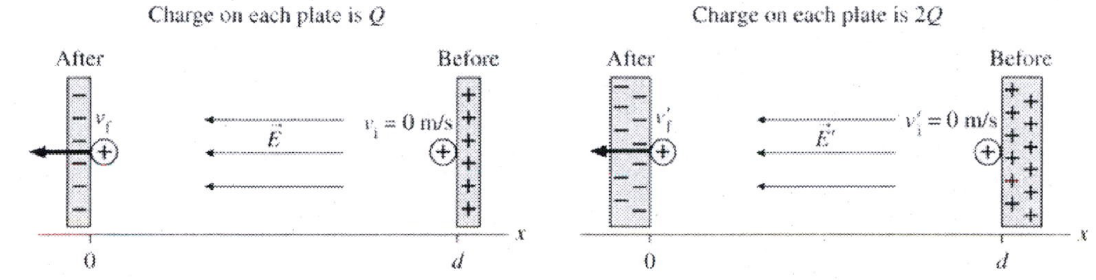

The figure shows the before-and-after pictorial representation.

Solve: The proton loses potential energy and gains kinetic energy as it moves toward the negative plate. The potential energy is defined as $U = U_0 + qEx$, where x is the distance from the negative plate and U_0 is the potential energy at the negative plate (at $x = 0$ m). Thus, the change in the potential energy of the proton is

$$\Delta U_p = U_f - U_i = (U_0 + 0\text{ J}) - (U_0 + qEd) = -qEd$$

The change in the kinetic energy of the proton is

$$\Delta K = K_f - K_i = \tfrac{1}{2}mv_f^2 - \tfrac{1}{2}mv_i^2 = \tfrac{1}{2}mv_f^2$$

Applying the law of conservation of energy $\Delta K + \Delta U_p = 0$ J, we have

$$\tfrac{1}{2}mv_f^2 + (-qEd) = 0\text{ J} \Rightarrow v_f^2 = \frac{2qEd}{m}$$

When the amount of charge on each plate is doubled, then the final velocity of the proton is

$$v_f'^2 = \frac{2qE'd}{m}$$

Dividing these equations,

$$\frac{v_f'^2}{v_f^2} = \frac{E'}{E} \Rightarrow v_f' = \sqrt{\frac{E'}{E}}v_f$$

For a parallel-plate capacitor $E = \eta/\varepsilon_0 = Q/A\varepsilon_0$. Therefore,

$$v_f' = \sqrt{\frac{Q'}{Q}}v_f = \sqrt{2}(50{,}000\text{ m/s}) = 70{,}711\text{ m/s}$$

Assess: The proton's velocity is expected to increase because an increased charge on the capacitor plates leads to a higher electric field between the plates and hence to an increased force on the proton.

29.7. Model: The charges are point charges.
Visualize: Please refer to Figure EX29.7.
Solve: For a system of point charges, the potential energy is the sum of the potential energies due to all distinct pairs of charges:

$$U_{\text{elec}} = \sum_{i,j} \frac{Kq_i q_j}{r_{ij}} = U_{12} + U_{13} + U_{23}$$

$$= \left(9.0\times10^9 \text{ N m}^2/\text{C}^2\right)\left[\frac{\left(3.0\times10^{-9}\text{ C}\right)\left(2.0\times10^{-9}\text{ C}\right)}{0.030\text{ m}} + \frac{\left(3.0\times10^{-9}\text{ C}\right)\left(2.0\times10^{-9}\text{ C}\right)}{0.040\text{ m}} + \frac{\left(2.0\times10^{-9}\text{ C}\right)\left(2.0\times10^{-9}\text{ C}\right)}{\sqrt{\left(0.030\text{ m}\right)^2+\left(0.040\text{ m}\right)^2}}\right]$$

$$= 1.80\times10^{-6}\text{ J} + 1.35\times10^{-6}\text{ J} + 0.72\times10^{-6}\text{ J} = 3.87\times10^{-6}\text{ J}$$

Assess: Note that $U_{12} = U_{21}$, $U_{13} = U_{31}$, and $U_{23} = U_{32}$.

29.9. Model: An external electric field supplies energy to a dipole.
Visualize: On an energy diagram, the oscillation occurs between the points where the potential-energy curve crosses the total energy line.

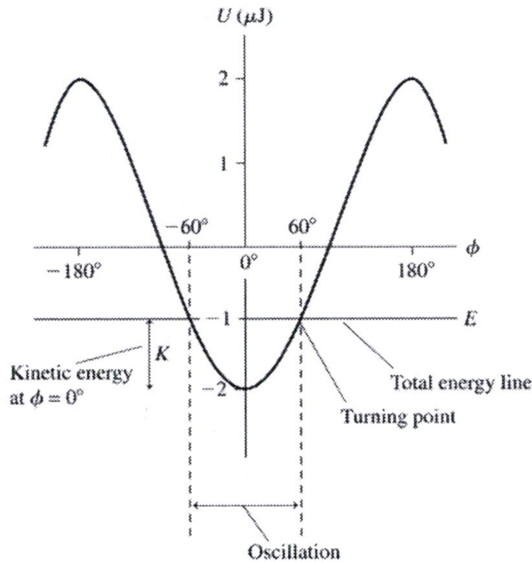

Solve: **(a)** The potential energy of an electric dipole moment in a uniform electric field is $U = -\vec{p}\cdot\vec{E} = -pE\cos\phi$. This means

$$U_{\phi=0^\circ} = -pE = -2\ \mu\text{J} \qquad U_{\phi=60^\circ} = -pE\cos 60^\circ = -\tfrac{1}{2}pE = -\tfrac{1}{2}(2\ \mu\text{J}) = -1\ \mu\text{J}$$

The mechanical energy $E_{\text{mech}} = U + K$. We know that at $\phi = 60°$, $K_{\phi=60^\circ} = 0$ J. So,

$$E_{\text{mech}} = U_{\phi=60^\circ} + K_{\phi=60^\circ} = -1\ \mu\text{J} + 0\text{ J} = -1\ \mu\text{J}$$

(b) Conservation of mechanical energy gives

$$U_{\phi=60^\circ} + K_{\phi=60^\circ} = U_{\phi=0^\circ} + K_{\iota=0^\circ} \Rightarrow -1\ \mu\text{J} + 0\text{ J} = -2\ \mu\text{J} + K_{\phi=0^\circ} \Rightarrow K_{\phi=0^\circ} = 1\ \mu\text{J}$$

29.11. Model: Energy is conserved. The potential energy is determined by the electric potential.
Visualize:

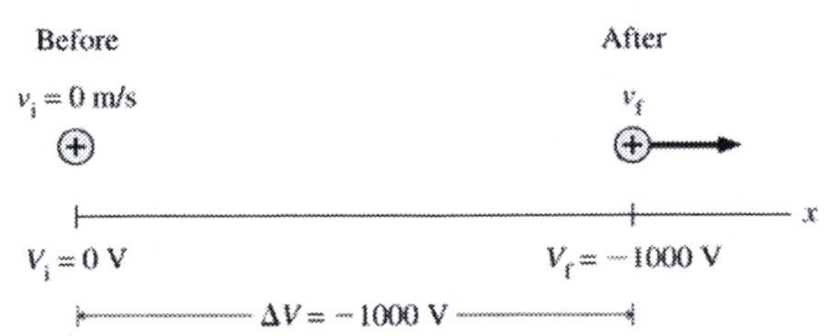

The figure shows a before-and-after pictorial representation of a proton moving through a potential difference of −1000 V. A positive charge *speeds up* as it moves into a region of lower potential ($U \rightarrow K$).

Solve: The potential energy of charge q is $U = qV$. Conservation of energy, expressed in terms of the electric potential V, is

$$K_f + qV_f = K_i + qV_i, \Rightarrow K_f = K_i + q(V_i - V_f) = 0\text{ J} - q\Delta V$$

$$\Rightarrow \tfrac{1}{2}mv_f^2 = -q(-1000\text{ V}) \Rightarrow v_f = \sqrt{\frac{2(1.60\times10^{-19}\text{ C})(1000\text{ V})}{1.67\times10^{-27}\text{ kg}}} = 4.38\times10^5\text{ m/s}$$

Assess: Note that the proton of our problem has nothing to do with creating the potential difference of −1000 V.

29.15. Model: Energy is conserved. The potential energy is determined by the electric potential.

Visualize:

Before

$v_i = 800{,}000$ m/s

V_i

After

$v_f = 0$ m/s

V_f

$\Delta V = V_f - V_i$

The figure shows a before-and-after pictorial representation of a proton moving through a potential difference.

Solve: **(a)** Because the proton is a positive charge and it slows down as it travels, it must be moving from a region of lower potential to a region of higher potential.

(b) Using the conservation of energy equation,

$$K_f + U_f = K_i + U_i \Rightarrow K_f + qV_f = K_i + qV_i$$

$$\Rightarrow V_f - V_i = \frac{1}{q}(K_i - K_f) = \frac{1}{(e)}\left(\tfrac{1}{2}mv_i^2 - 0\text{ J}\right)$$

$$\Rightarrow \Delta V = \frac{mv_i^2}{2e} = \frac{(1.67\times10^{-27}\text{ kg})(800{,}000\text{ m/s})^2}{2(1.60\times10^{-19}\text{ C})} = 3340\text{ V}$$

Assess: A positive ΔV confirms that the proton moves into a higher potential region.

29.19. Model: The electric potential difference between the plates is determined by the uniform electric field in the parallel-plate capacitor.

Solve: **(a)** The potential difference ΔV_C across a capacitor of spacing d is related to the electric field inside by

$$E = \frac{\Delta V_C}{d} \Rightarrow \Delta V_C = Ed = (1.0\times10^5\text{ V/m})(0.0020\text{ m}) = 200\text{ V}$$

(b) The electric field of a capacitor is related to the surface charge density by

$$E = \frac{\eta}{\varepsilon_0} = \frac{Q/A}{\varepsilon_0}$$

$$\Rightarrow Q = \varepsilon_0 AE = (8.85\times10^{-12}\text{ C}^2\text{/N m}^2)(4.0\times10^{-4}\text{ m}^2)(1.0\times10^5\text{ V/m}) = 3.5\times10^{-10}\text{ C}$$

29.23. Model: The charge is a point charge.

Visualize: Please refer to Figure EX29.23.

Solve: **(a)** The electric potential of the point charge q is

$$V = \frac{1}{4\pi\varepsilon_0}\frac{q}{r} = (9.0\times10^9\text{ N m}^2\text{/C}^2)\left(\frac{2.0\times10^{-9}\text{ C}}{r}\right) = \frac{18.0\text{ N m}^2\text{/C}}{r}$$

For points A and B, $r = 0.010$ m. Thus,

$$V_A = V_B = \frac{18.0\text{ N m}^2\text{/C}}{0.010\text{ m}} = 1800\ \frac{\text{Nm}}{\text{C}} = 1800\left(\frac{\text{V}}{\text{m}}\right)\text{m} = 1.80\text{ kV}$$

For point C, $r = 0.020$ m and $V_C = 900$ V.

(b) The potential differences are

$$\Delta V_{AB} = V_B - V_A = 1.80 \text{ kV} - 1.80 \text{ kV} = 0 \text{ V} \qquad \Delta V_{BC} = V_C - V_B = 0.90 \text{ kV} - 1.80 \text{ kV} = -0.90 \text{ kV}$$

29.29. Model: The net potential is the sum of the scalar potentials due to each charge.
Visualize:

$q_1 = +3.0$ nC $\qquad q_2 = -1.0$ nC, x (cm): 0, 1, 2, 3, 4

Solve: Let the point on the x-axis where the electric potential is zero be at a distance x from the origin. At this point, $V_1 + V_2 = 0$ V. This means

$$\frac{1}{4\pi\varepsilon_0}\left\{\frac{3.0\times10^{-9}\text{ C}}{x}+\frac{-1.0\times10^{-9}\text{ C}}{|x-4.0\text{ cm}|}\right\}=0\text{ V}\Rightarrow -x+3|x-4.0\text{ cm}|=0\text{ cm}$$

Either $-x+3(x-4.0\text{ cm})=0$ cm, or $-x+3(4.0\text{ cm}-x)=0$ cm. In the first case, x = 6.0 cm and, in the second case, x = 3 cm. That is, we have two points on the x-axis where the potential is zero.

29.31. Model: While the potential is the sum of the scalar potentials due to each charge, the electric field is the vector sum of the electric fields due to each charge.
Visualize: Please refer to Figure EX29.31.
Solve: **(a)** E_x points left for $x > b$ because E_x is negative. So, the charge at $x = b$ is negative. For $x < a$, E_x points left, so the charge at $x = a$ is positive. For $a < x < b$, E_x is positive which is consistent with the charge choices.

(b) By the symmetry of the drawing about the middle, it appears that the magnitudes of the charges are the same. Thus $\left|q_a/q_b\right| = 1$.

(c)

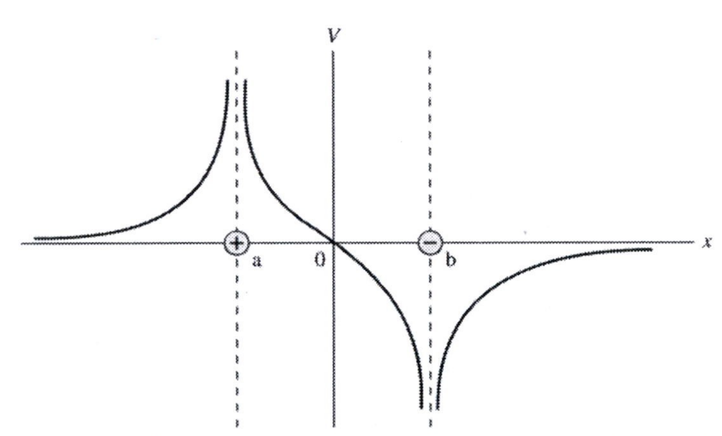

The graph of the electric potential shown in the figure is consistent with the electric field as well as the charges.

29.37. Model: While the net potential is the sum of the potentials due to each charge, the net electric field is the vector sum of the electric fields.
Visualize:

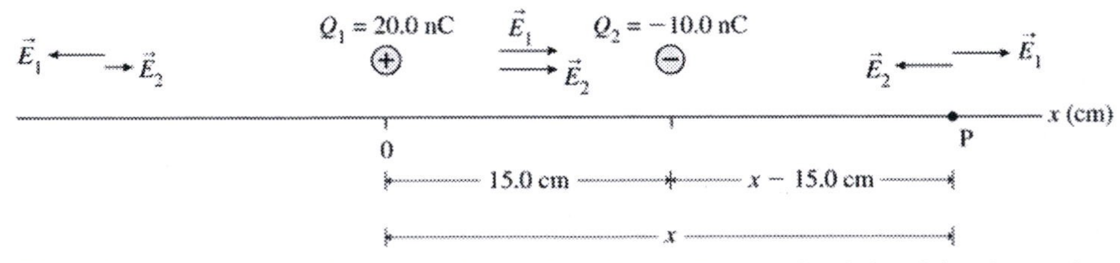

The charge $Q_1 = 20.0$ nC is at the origin. The charge $Q_2 = -10.0$ nC is 15.0 cm to the right of the charge Q_1 on the x-axis.

Solve: **(a)** As the pictorial representation shows, the point P on the x-axis where the electric field is zero can only be on the right side of the charge Q_2, that is, at $x \geq 15.0$ cm. At this point $E_1 = E_2$, so we have

$$\frac{1}{4\pi\varepsilon_0}\frac{20.0\times10^{-9}\text{ C}}{x^2} = \frac{1}{4\pi\varepsilon_0}\frac{10.0\times10^{-9}\text{ C}}{(x-15.0\text{ cm})^2} \Rightarrow x^2 = 2(x-15.0\text{ cm})^2$$

$$\Rightarrow x^2 - (60.0\text{ cm})x + 450\text{ cm}^2 = 0 \Rightarrow x = \frac{(60.0\text{ cm}) \pm \sqrt{3600\text{ cm}^2 - 1800\text{ cm}^2}}{2}$$

$$\Rightarrow x = 51.2\text{ cm and }8.8\text{ cm.}$$

The root $x = 8.8$ cm is not possible physically. So, the electric fields cancel out at $x = 51.2$ cm. The electric potential at this point is

$$V = \frac{1}{4\pi\varepsilon_0}\frac{Q_1}{r_1} + \frac{1}{4\pi\varepsilon_0}\frac{Q_2}{r_2} = (9.0\times10^9\text{ N m}^2/\text{C}^2)\left[\frac{20.0\times10^{-9}\text{ C}}{0.512\text{ m}} + \frac{-10.0\times10^{-9}\text{ C}}{(0.512\text{ m} - 0.150\text{ m})}\right] = +103\text{ V}$$

(b) The point on the x-axis where the potential is zero can be obtained from the condition $V_1 + V_2 = 0$ V, which is

$$\frac{1}{4\pi\varepsilon_0}\frac{Q_1}{r_1} + \frac{1}{4\pi\varepsilon_0}\frac{Q_2}{r_2} = 0\text{ V} \Rightarrow \frac{20.0\times10^{-9}\text{ C}}{x} + \frac{-10.0\times10^{-9}\text{ C}}{(15.0\text{ cm} - x)} = 0$$

$$\Rightarrow 2(15.0\text{ cm} - x) - x = 0 \Rightarrow x = 10.0\text{ cm}$$

The electric field 10.0 cm away from charge Q_1 is

$$\vec{E}_{\text{net}} = \vec{E}_1 + \vec{E}_2 = \frac{1}{4\pi\varepsilon_0}\frac{20.0\times10^{-9}\text{ C}}{(0.100\text{ m})^2}\hat{i} + \frac{1}{4\pi\varepsilon_0}\frac{(10.0\times10^{-9}\text{ C})}{(0.050\text{ m})^2}\hat{i} = 5.40\times10^4\hat{i}\text{ N/C}$$

29.39. Model: The charged beads are point charges.
Visualize:

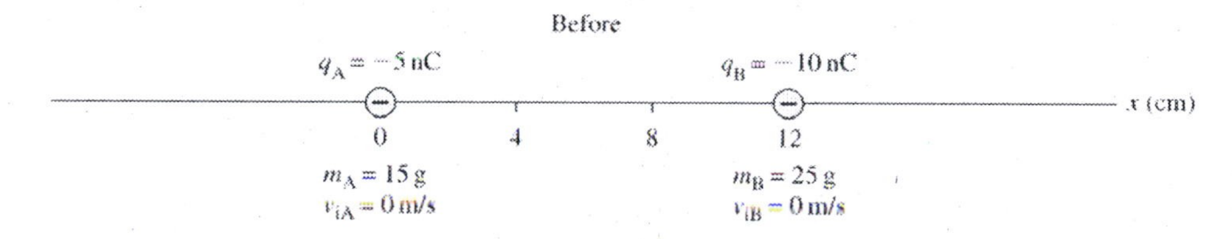

Solve: To find the maximum speeds v_{fA} and v_{fB} it is easier to use the conservation of momentum and conservation of energy equations than kinematics, as suggested in the hint. The beads achieve their maximum speeds when they are infinitely separated. The momentum conservation equation along x-direction $p_{\text{after}} = p_{\text{before}} = 0$ kg m/s means

$$-m_A v_{fA} + m_B v_{fB} = 0\text{ kg m/s} \Rightarrow -(0.015\text{ kg})v_{fA} + (0.025\text{ kg})v_{fB} = 0\text{ kg m/s} = 0\text{ kg m/s} \Rightarrow v_{fB} = \tfrac{3}{5}v_{fA}$$

The conservation of energy conservation is $U_f + K_f = U_i + K_i$. Noting that $U_f \to 0$ J as the masses are infinitely separated and

$$U_i = \frac{1}{4\pi\varepsilon_0}\frac{Q_A Q_B}{r_{AB}} = \frac{(9.0\times10^9\text{ N m}^2/\text{C}^2)(5.0\times10^{-9}\text{ C})(10.0\times10^{-9}\text{ C})}{0.12\text{ m}} = 3.75\times10^{-6}\text{ J}$$

$$\Rightarrow 0\text{ J} + \left(\tfrac{1}{2}m_A v_{fA}^2 + \tfrac{1}{2}m_B v_{fB}^2\right) = 3.75\times10^{-6}\text{ J} + 0\text{ J}$$

$$\Rightarrow \tfrac{1}{2}(0.015\text{ kg})v_{fA}^2 + \tfrac{1}{2}(0.025\text{ kg})\left(\tfrac{3}{5}v_{fA}\right)^2 = 3.75\times10^{-6}\text{ J} \Rightarrow 0.012v_{fA}^2 = 3.75\times10^{-6}\text{ J}$$

Solving these equations yields $v_{fA} = 1.77$ cm/s and $v_{fB} = \frac{3}{5}(1.77 \text{ cm/s}) = 1.06$ cm/s. These are the maximum speeds of the two beads and occur when they are infinitely separated from each other.

29.43. Model: Energy is conserved.
Solve: (a)

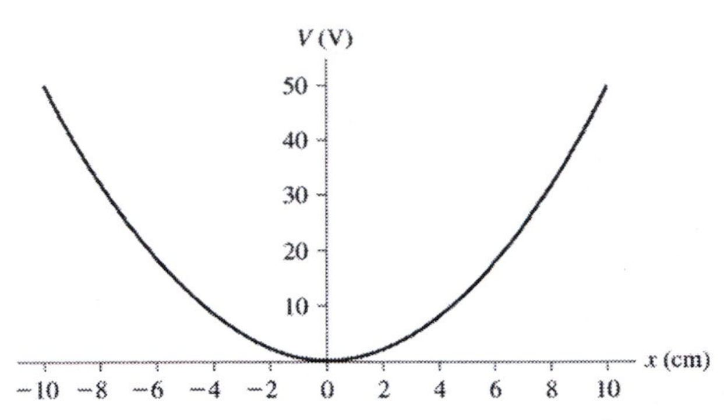

(b) The potential energy of the positive charge as a function of x is $U(x) = 5000\,qx^2$. This is analogous to the potential energy of a mass on a spring. Thus the motion is simple harmonic motion.
(c) Turning points at ±8.0 cm mean that $x_{max} = \pm 8.0$ cm. At the turning points, the energy is all potential energy and the kinetic energy is zero. The mechanical energy of the charged particle is

$$E = U + K = U_{max} + 0 \text{ J} = 5000\,q(x_{max})^2 = 5000(10.0\times10^{-9} \text{ C})(0.080 \text{ m})^2 = 3.2\times10^{-7} \text{ J}$$

(d) The conservation of energy equation $K_1 + U_1 = K_2 + U_2$ is

$$K_{max} + 0 \text{ J} = 0 \text{ J} + U_{max} \Rightarrow \tfrac{1}{2}mv_{max}^2 = 3.20\times10^{-7} \text{ J} \Rightarrow v_{max} = \sqrt{\frac{2(3.20\times10^{-7} \text{ J})}{(1.0\times10^{-3} \text{ kg})}} = 2.5 \text{ cm/s}$$

29.53. Model: Outside a charged sphere the electric potential is identical to that of a point charge at the center.
Visualize:

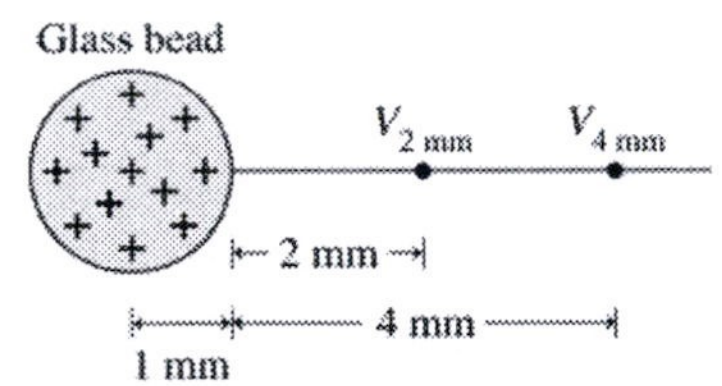

Solve: For $r \geq R$, $V = Q/4\pi\varepsilon_0 r$. The potentials at the two points are

$$V_{2\text{ mm}} = \frac{(9.0\times10^9 \text{ N m}^2/\text{C}^2)Q}{(1 \text{ mm} + 2 \text{ mm})} = \frac{(9.0\times10^9 \text{ N m}^2/\text{C}^2)Q}{3.0\times10^{-3} \text{ m}} \qquad V_{4\text{ mm}} = \frac{(9.0\times10^9 \text{ N m}^2/\text{C}^2)Q}{5.0\times10^{-3} \text{ m}}$$

$$\Rightarrow V_{2\text{ mm}} - V_{4\text{ mm}} = +500 \text{ V} = (9.0\times10^9 \text{ N m}^2/\text{C}^2)Q\left(\frac{1}{3.0\times10^{-3} \text{ m}} - \frac{1}{5.0\times10^{-3} \text{ m}}\right)$$

$$\Rightarrow Q = \frac{500 \text{ V}}{(9.0\times10^9 \text{ N m}^2/\text{C}^2)}\left(\frac{15.0\times10^{-6} \text{ m}^2}{2.0\times10^{-3} \text{ m}}\right) = 4.2\times10^{-10} \text{ C}$$

Assess: Do not forget to include the radius of the glass bead in r.

29.55. Model: Energy is conserved. Because the mercury nucleus is very large compared to the proton, we will assume that the nucleus does not move (no recoil) and that the proton is essentially a point particle with no diameter.
Visualize:

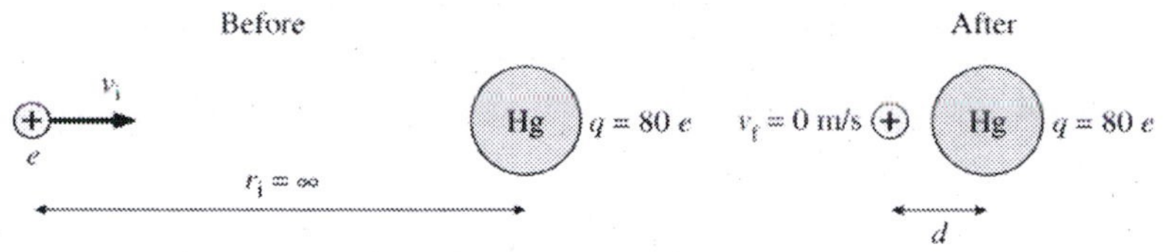

Before

v_i e Hg $q = 80\,e$ $r_i = \infty$

After

$v_f = 0$ m/s Hg $q = 80\,e$ d

The proton is fired from a distance much greater than the nuclear diameter, so $r_i = \infty$ and $U_i = 0$ J. Because the nucleus is so small, a proton that is even a few atoms away is, for all practical purposes, at infinity. As the proton approaches the nucleus, it is slowed by the repulsive electric force. At the end point of the problem, the proton reaches the distance of closest approach (d) with $v_f = 0$ m/s. (The proton won't remain at this point but will be pushed back out again, but the subsequent motion is not part of the problem.)
Solve: Initially, the proton has kinetic energy but no potential energy. At the point of closest approach, where $v_f =$ 0 m/s, the proton has potential energy but no kinetic energy. Energy is conserved, so $K_f + U_f = K_i + U_i$. This equation is

$$0\text{ J} + \frac{e(80e)}{4\pi\varepsilon_0 d} = \tfrac{1}{2}m_{\text{proton}}v_i^2 + 0\text{ J}$$

$$\Rightarrow d = \frac{160e^2}{4\pi\varepsilon_0 m_{\text{proton}} v_i^2} = \frac{(9.0\times10^9\text{ N m}^2/\text{C}^2)(160)(1.6\times10^{-19}\text{ C})^2}{(1.67\times10^{-27}\text{ kg})(4.0\times10^7\text{ m/s})^2} = 1.38\times10^{-14}\text{ m} = 13.8\text{ fm}$$

The radius of the nucleus is 7.0 fm, so the proton's closest approach to the surface is 6.8 fm.

29.59. Model: The electric field inside a capacitor is uniform.
Solve: (a) Because the parallel-plate capacitor was connected to the terminals of a 15 V battery for a long time, the potential difference across the capacitor right after the battery is disconnected is ΔV= 15 V. The electric field strength inside the capacitor is

$$E = \frac{\Delta V_C}{d} = \frac{15\text{ V}}{0.50\times10^{-2}\text{ m}} = 3000\text{ V/m} = 3.0\text{ kV/m}$$

Because $E = \eta/\varepsilon_0$ for a parallel-plate capacitor and $\eta = Q/A$, the total charge on each plate is

$$Q = EA\varepsilon_0 = (3000\text{ V/m})\pi(0.050\text{ m})^2(8.85\times10^{-12}\text{ C}^2/\text{Nm}^2) = 2.1\times10^{-10}\text{ C}$$

(b) After the electrodes are pulled away to a separation of $d' = 1.0$ cm, the charges on the plates are unchanged. That is, $Q' = Q$. Because $A' = A$, the electric field inside the capacitor is also unchanged. So, $E' = E$. The potential difference across the capacitor is $\Delta V_C' = E'd'$. Because d increases from 0.50 cm to 1.0 cm ($d' = 1.0$ cm), the potential difference $\Delta V_C'$ increases from 15 V to 30 V.
(c) When the electrodes are expanded, the new area is $A' = \pi(r')^2 = \pi(2r)^2 = 4A$. The charge Q' on the capacitor plates, however, stays the same as before ($Q' = Q$). The electric field is

$$E' = \frac{\eta'}{\varepsilon_0} = \frac{Q'}{A'\varepsilon_0} = \frac{Q}{4A\varepsilon_0} = \frac{E}{4} = \frac{3000\text{ V/m}}{4} = 750\text{ V/m} = 0.75\text{ kV/m}$$

The potential difference across the capacitor plates $\Delta V_C' = E'd' = (750\text{ V})(0.050\text{ m}) = 3.8$ V.

29.63. Solve: (a) Because the excess charge resides on the outer surface of a conductor, the charge placed on the inside surface of a hollow metal sphere will move rapidly to the outside surface.
(b) Because a spherical shell of charge has the same electric potential as a point charge Q at the center, the potential on the surface is

$$V = \frac{1}{4\pi\varepsilon_0}\frac{Q}{R} \Rightarrow 500{,}000\text{ V} = (9.0\times10^9\text{ N m}^2/\text{C}^2)\left(\frac{Q}{0.15\text{ m}}\right) \Rightarrow Q = 8.3\,\mu\text{C}$$

(c) The electric field inside a conductor is zero, so the electric field just inside the sphere is zero. From Gauss's Law, the electric field outside a charged conductor is

$$E=\frac{\eta}{\varepsilon_0}=\frac{Q}{A\varepsilon_0}=\frac{8.33\times10^{-6}\text{ C}}{4\pi(0.15\text{ m})^2(8.85\times10^{-12}\text{ C}^2/\text{Nm}^2)}=3.3\times10^6\text{ V/m}$$

29.65. Model: The potential at any point is the superposition of the potentials due to all charges. Outside a uniformly charged sphere, the electric potential is identical to that of a point charge Q at the center.
Visualize: Please refer to Figure P29.65. Sphere A is the sphere on the left and sphere B is the one on the right.
Solve: The potential at point a is the sum of the potentials due to the spheres A and B:

$$V_{\text{a}}=V_{\text{A at a}}+V_{\text{B at a}}=\frac{1}{4\pi\varepsilon_0}\frac{Q_{\text{A}}}{R_{\text{A}}}+\frac{1}{4\pi\varepsilon_0}\frac{Q_{\text{B}}}{0.70\text{ m}}$$
$$=(9.0\times10^9\text{N m}^2/\text{C}^2)\frac{100\times10^{-9}\text{ C}}{0.30\text{ m}}+(9.0\times10^9\text{ N m}^2/\text{C}^2)\frac{25\times10^{-9}\text{ C}}{0.70\text{ m}}$$
$$=3000\text{ V}+321\text{ V}=3321\text{ V}$$

Similarly, the potential at point b is the sum of the potentials due to the spheres A and B:

$$V_{\text{b}}=V_{\text{B at b}}+V_{\text{A at b}}=\frac{1}{4\pi\varepsilon_0}\frac{Q_{\text{B}}}{R_{\text{B}}}+\frac{1}{4\pi\varepsilon_0}\frac{Q_{\text{A}}}{0.95\text{ m}}$$
$$=(9.0\times10^9\text{ N m}^2/\text{C}^2)\left(\frac{25\times10^{-9}\text{ C}}{0.05\text{ m}}+\frac{100\times10^{-9}\text{ C}}{0.95\text{ m}}\right)$$
$$=4500\text{ V}+947\text{ V}=5447\text{ V}$$

Thus, the potential at point b is higher than the potential at a. The difference in potential is $V_{\text{b}}-V_{\text{a}}=5447$ V – 3321 V = 2126 V = 2.1 kV.
Assess: $V_{\text{A at a}}=3000$ V and the sphere B has a potential of 225 V at point a. The spherical symmetry dictates that the potential on a sphere's surface be the same everywhere. So, in calculating the potential at point a due to the sphere B we used the center-to-center separation of 1.0 m rather than a separation of 100 cm – 30 cm = 70 cm from the center of sphere B to the point a. The former choice leads to the same potential everywhere on the surface whereas the latter choice will lead to a distribution of potentials depending upon the location of the point a. Similar reasoning also applies to the potential at point b.

29.71. Model: Because the rod is thin, assume the charge lies along the semicircle of radius R.
Visualize: Please refer to Figure P29.71. The bent rod lies in the xy-plane with point P as the center of the semicircle. Divide the semicircle into N small segments of length Δs and of charge $\Delta Q=(Q/\pi R)\Delta s$, each of which can be modeled as a point charge. The potential V at P is the sum of the potentials due to each segment of charge.
Solve: The total potential is

$$V=\sum_i V_i=\sum_i\frac{1}{4\pi\varepsilon_0}\frac{\Delta Q}{R}=\frac{1}{4\pi\varepsilon_0 R}\sum_i\left(\frac{Q}{\pi R}\right)\Delta s=\frac{1}{4\pi\varepsilon_0}\frac{Q}{\pi R^2}\sum_i R\Delta\theta=\frac{1}{4\pi\varepsilon_0}\frac{Q}{\pi R}\sum_i\Delta\theta.$$

All of the terms come to the front of the summation because these quantities did not change as far as the summation is concerned. The summation does not have to convert to an integral because the sum of all the $\Delta\theta$ around the semicircle is π. Hence, the potential at the center of a charged semicircle is

$$V_{\text{center}}=\frac{1}{4\pi\varepsilon_0}\frac{Q\pi}{\pi R}=\frac{1}{4\pi\varepsilon_0}\frac{Q}{R}$$

29.75. Solve: (a) Two 3.0 nC charges are distance d apart. At a point 3.0 cm from one charge, on the side opposite the other charge, the potential is 1200 V. Find the separation d between the charges.
(b) The given equation simplifies to

$$900\text{ N m/C}+\frac{27\text{ N m}^2/\text{C}}{0.030\text{ m}+d}=1200\text{ V}\Rightarrow\frac{27\text{ N m}^2/\text{C}}{0.030\text{ m}+d}=300\text{ V}\Rightarrow d=0.060\text{ m}=6.0\text{ cm}$$

30

POTENTIAL AND FIELD

Exercises and Problems

30.3. Model: The potential varies along the x-axis.
Visualize: Please refer to Figure EX30.3.
Solve: The connection between the potential difference and the electric field is

$$\Delta V = -\int_{i}^{f} E_x\, dx = -(\text{area under the } E_x\text{-versus-}x \text{ curve between } x_i \text{ and } x_f.)$$

The graph from $x_i = 1.0$ m and $x_f = 3.0$ m is a straight line. The area under it is that of a triangle:

$$\Delta V = -\frac{1}{2}\times 2.0 \text{ m} \times 200 \text{ V/m} = -200 \text{ V}$$

Assess: E_x is positive in this region, meaning that the field points in the $+x$-direction. Because the electric field points "downhill," in the direction of decreasing potential, we expected ΔV to be negative.

30.7. Solve: The emf is defined as the work done per unit charge by the charge escalator or the battery. That is,

$$\mathcal{E} = \frac{W_{\text{chem}}}{q} = \frac{0.60 \text{ J}}{0.050 \text{ C}} = 12 \text{ V}$$

30.11. Model: The electric field is the negative of the slope of the graph of the potential function.
Visualize: Please refer to Figure EX30.11.
Solve: There are three regions of different slope. For 0 cm $< x <$ 10 cm and 20 cm $< x <$ 30 cm,

$$\frac{\Delta V}{\Delta x} = 0 \text{ V/m} \Rightarrow E_x = 0 \text{ V/m}$$

For 10 cm $< x <$ 20 cm,

$$\frac{\Delta V}{\Delta x} = \frac{-100 \text{ V} - (100 \text{ V})}{0.20 \text{ m} - 0.10 \text{ m}} = -2000 \text{ V/m} \Rightarrow E_x = +2000 \text{ V/m}$$

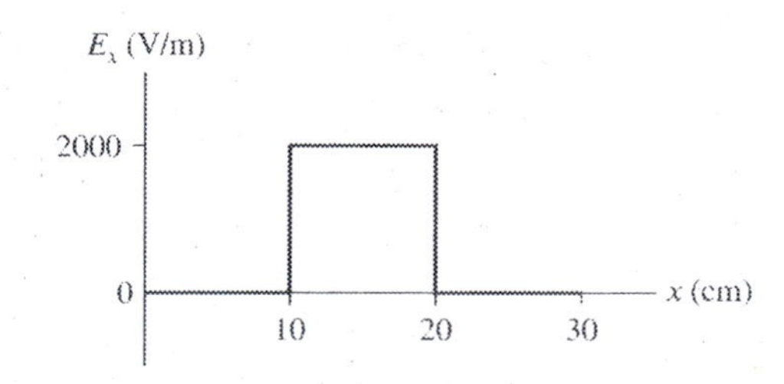

Assess: Because $E_s = -dV/ds$, the electric field is zero where the potential is not changing.

30.13. Visualize:

$V = -1000$ V $\quad E_x \quad V = +1000$ V $\quad x$ (m)

$-1.0 \quad 0 \quad +1.0$

i → f

f ← i

Solve: The electric potential difference ΔV between two points in a uniform electric field is

$$V(x_f) - V(x_i) = -\int E_x dx = -E_x(x_f - x_i)$$

Choosing $x_i = -1.0$ m and $x_f = +1.0$ m,

$$+1000 \text{ V} - (-1000 \text{ V}) = -E_x[1.0 \text{ m} - (-1.0 \text{ m})] \Rightarrow E_x = -1.0 \text{ kV/m}$$

Alternatively, $x_i = 1.0$ m and $x_f = -1.0$ m. For this choice,

$$-1000 \text{ V} - (+1000 \text{ V}) = -E_x[-1.0 \text{ m} - (1.0 \text{ m})] \Rightarrow E_x = -1.0 \text{ kV/m}$$

Assess: The choice of initial and final positions does not change the physical nature of the electric field or the potential difference.

30.15. Solve: **(a)** Since $E_x = -\dfrac{dV}{dx}$, we have

$$E_x = -\frac{d}{dx}(50x - 100/x) \text{ V/m} = -(50 + 100/x^2) \text{ V/m}$$

At $x = 1.0$ m,

$$E_x = -(50 + 100/(1.0)^2) \text{ V/m} = -150 \text{ V/m}$$

(b) At $x = 2.0$ m,

$$E_x = -\left(50 + \frac{100}{(2.0)^2}\right) \text{ V/m} = -75 \text{ V/m}$$

30.23. Solve: According to Equation 30.23,

$$C_{eq} = \left(\frac{1}{C_1} + \frac{1}{C_2} + \frac{1}{C_3}\right)^{-1} = \left(\frac{1}{6\ \mu\text{F}} + \frac{1}{10\ \mu\text{F}} + \frac{1}{16\ \mu\text{F}}\right)^{-1} = 3.0 \times 10^{-6} \text{ F} = 3.0\ \mu\text{F}$$

30.27. Model: Assume the battery is ideal.
Visualize: Please refer to Figure EX30.27.
Solve: For an ideal battery, the potential difference across the capacitor is the same as the emf of the battery. Thus,

$$\mathcal{E} = \Delta V_C = \frac{Q}{C} \Rightarrow Q = C\mathcal{E} = (10 \times 10^{-6} \text{ F})(1.5 \text{ V}) = 15.0 \times 10^{-6} \text{ C}$$

30.29. Solve: The graph of Figure EX30.29 shows that the capacitor is fully charged at 3.0 s and that its charge is 200 μC. From t = 0 s to t = 3.0 s, the charge on the capacitor follows the equation $Q = (200\ \mu\text{C}/3.0 \text{ s})t$. Using Equation 30.26, the energy stored in the capacitor as a function of time is

$$U_C = \frac{1}{2}\frac{Q^2}{C} = \frac{1}{2}\left(\frac{200\ \mu\text{C}}{3.0 \text{ s}}\right)^2 \frac{t^2}{2.0\ \mu\text{F}} = (1.111 \times 10^{-3} \text{ C}^2/\text{s F})t^2$$

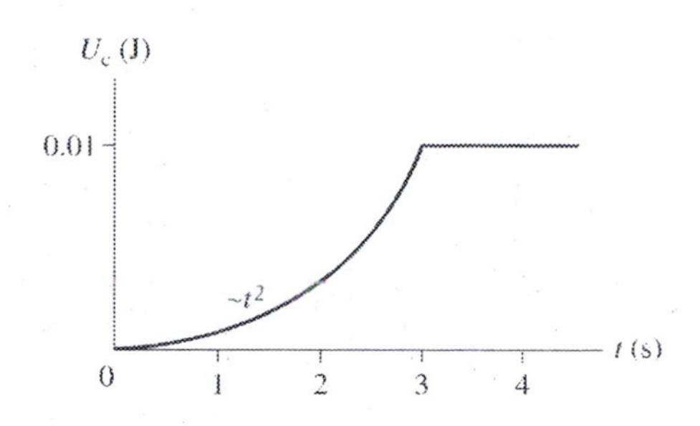

30.35. Model: The disks form a parallel-plate capacitor.
Solve: **(a)** The charge on each of the metal disks is

$$Q = C\Delta V_C = \kappa_{\text{Pyrex}} C_0 \Delta V_C = \kappa_{\text{Pyrex}} \frac{\varepsilon_0 A}{d} \Delta V_C$$

The charge density on a disk is

$$\eta = \frac{Q}{A} = \kappa_{\text{Pyrex}} \frac{\varepsilon_0 \Delta V_C}{d} = (4.7)\frac{(8.85\times10^{-12}\text{ C}^2/\text{N m}^2)(1000\text{ V})}{(0.50\times10^{-3}\text{ m})}$$
$$= 8.3\times10^{-5}\text{ C/m}^2 = 83\ \mu\text{C/m}^2$$

(b) In the absence of the dielectric, the surface charge density on the disks is

$$\eta_0 = \frac{\eta}{\kappa_{\text{Pyrex}}} = 1.77\times10^{-5}\text{ C/m}^2.$$

The surface charge density on the Pyrex is

$$\eta_{\text{induced}} = \eta_0\left(1 - \frac{1}{\kappa_{\text{Pyrex}}}\right) = (1.77\times10^{-5}\text{ C/m}^2)\left(1 - \frac{1}{4.7}\right)$$
$$= 1.39\times10^{-5}\text{ C/m}^2 = 13.9\ \mu\text{C/m}^2$$

Assess: The induced charge density on the dielectric is less than the charge density on the plates.

30.37. Solve: (a)

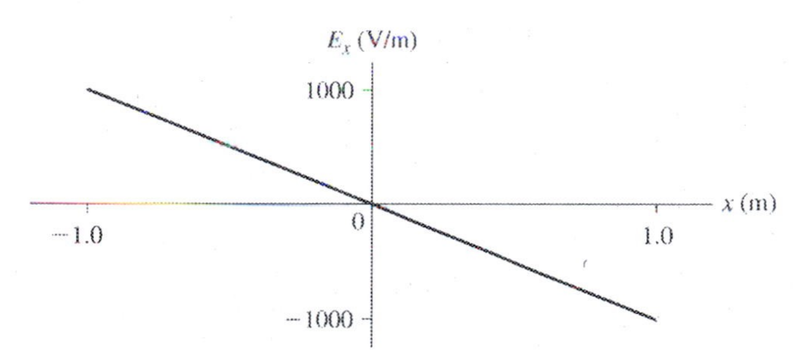

(b) Equation 30.3 gives the potential difference between two points in space:

$$\Delta V = -\int_{x_i}^{x_f} E_x\, dx = -\int_{-20\text{ cm}}^{30\text{ cm}} (-1000x\text{ V/m})\, dx$$
$$= +\left[1000\frac{x^2}{2}\right]_{-0.20\text{ m}}^{0.30\text{ m}} \text{V} = +\frac{1000}{2}\left[(0.30\text{ m})^2 - (-0.20\text{ m})^2\right]\text{V} = +65\text{ V}$$

Assess: E is positive for negative x but negative for positive x, so the potential difference depends on the square of the positions.

30.39. Visualize:

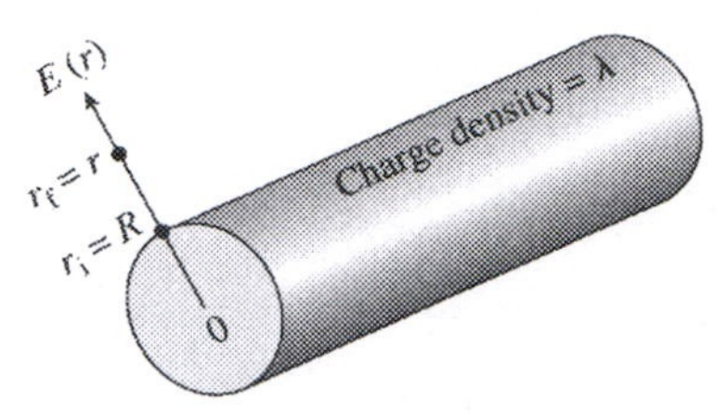

Solve: Equation 30.3 gives the potential difference between two points in space

$$\Delta V = V(r_f) - V(r_i) = -\int_{r_i}^{r_f} E_r dr$$

$$\Rightarrow V(r) - V(R) = -\int_R^r \frac{\lambda}{2\pi\varepsilon_0 r} dr = -\frac{\lambda}{2\pi\varepsilon_0}[\ln r]_R^r = -\frac{\lambda}{2\pi\varepsilon_0}[\ln r - \ln R] = -\frac{\lambda}{2\pi\varepsilon}\ln\frac{r}{R}$$

$$\Rightarrow V(r) = V_0 - \frac{\lambda}{2\pi\varepsilon_0}\ln\frac{r}{R}$$

Assess: At $r = R$, $\ln(R/R) = 0$ and $V(R) = V_0$, as it must be due to the assumed constant of integration.

30.41. Model: Assume the electrodes form parallel-plate capacitors with a uniform electric field between the plates.
Visualize:

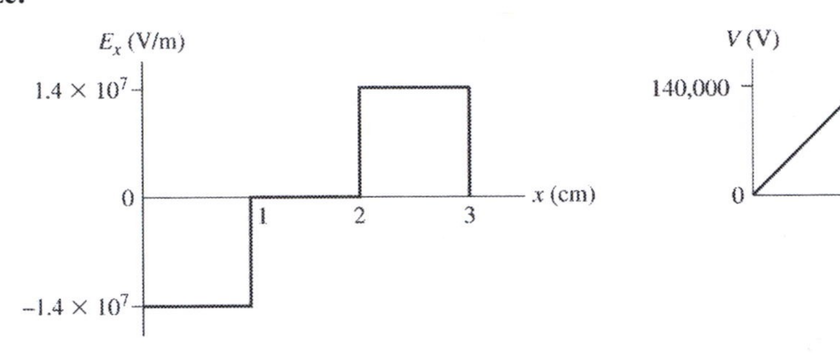

Please refer to Figure P30.41. The three metal electrodes serve as plates for two capacitors. On the middle electrode, half the charge is located on the left face and half on the right face, thus forming two capacitors. Each plate of the two capacitors carries a charge of ±50 nC.
Solve: **(a)** In the space 0 cm $< x <$ 1 cm, the electric field points to the left and its magnitude is

$$E = \frac{\eta}{\varepsilon_0} = \frac{Q}{A\varepsilon_0} = \frac{50\times10^{-9}\text{ C}}{(0.02\text{ m})^2(8.85\times10^{-12}\text{ C}^2/\text{N m}^2)} = 1.41\times10^7\text{ V/m}$$

In the region 1 cm $\le x \le$ 2 cm, $\vec{E} = 0$ because in electrostatics the inside of a conductor has no free charge. The electric field in the region 2 cm $< x <$ 3 cm points to the right and has the same magnitude as the electric field in the region 0 cm $< x <$ 1 cm.
(b) The potential difference between two points in space with a uniform electric field is

$$\Delta V = V_f - V_i = E(x_f - x_i)$$

Assuming that the negative plate at $x = 0$ m is at zero potential ($V_i = 0$ V at $x_i = 0$ cm), $V_f = x_f E$, or simply $V = xE$. Thus, the potential increases linearly with distance x from the negative plate in the region $0 \le x \le 1$. At $x = 1$ cm, the potential is

$$V = xE = (1.0\times10^{-2}\text{ m})(1.41\times10^7\text{ V/m}) = 1.41\times10^{-5}\text{ V}$$

The potential must be the same throughout the region 1 cm $\le x \le$ 2 cm. If this were not the case, we would not have an electrostatic situation with the electric field $E = 0$ V/m. Using the previous reasoning, the potential decreases linearly in the region 2 cm $< x <$ 3 cm.

30.43. Model: The electric field is the negative of the slope of the potential function.
Solve: The on-axis potential of a charged disk was obtained in Chapter 29 to be

$$V_{\text{disk}} = \frac{Q}{2\pi\varepsilon_0 R^2}\left[\sqrt{R^2+z^2} - z\right]$$

where the charge on the disk of radius R is Q and the point is a distance z away from the center of the disk. Because $E = -dV/dz$, we have

$$E_{\text{disk}} = -\frac{dV_{\text{disk}}}{dz} = -\frac{Q}{2\pi\varepsilon_0 R^2}\left[\frac{\frac{1}{2}(2z)}{\sqrt{R^2+z^2}} - 1\right] = \frac{Q}{2\pi\varepsilon_0 R^2}\left[1 - \frac{z}{\sqrt{R^2+z^2}}\right]$$

Assess: Using binomial expansion with $z >> R$, we have

$$E_{\text{disk}} = \frac{Q}{2\pi\varepsilon_0 R^2}\left[1 - \frac{z}{z\left(1+R^2/z^2\right)^{1/2}}\right] = \frac{Q}{2\pi\varepsilon_0 R^2}\left[1-\left(1+R^2/z^2\right)^{-1/2}\right] = \frac{Q}{2\pi\varepsilon_0 R^2}\left[1-1+\tfrac{1}{2}R^2/z^2+\cdots\right] = \frac{Q}{4\pi\varepsilon_0 z^2}$$

That is, the disk behaves like a point charge. This is an expected result.

30.47. Model: The electric field is the negative of the slope of the graph of the potential function.

Solve: The electric potential in a region of space is $V = (150\,x^2 - 200\,y^2)$ V where x and y are in meters. The x- and y-components are

$$E_x = -\frac{dV}{dx} = -(300\,x)\text{ V/m} \qquad E_y = -\frac{dV}{dy} = +(400\,y)\text{ V/m}$$

At $(x,y) = (2.0\text{ m}, 2.0\text{ m})$, $E_x = -600$ V/m and $E_y = 800$ V/m. The magnitude and direction of the electric field are

$$E = \sqrt{E_x^2 + E_y^2} = \sqrt{(-600\text{ V/m})^2 + (800\text{ V/m})^2} = 1000\text{ V/m}$$

$$\tan\theta = \frac{E_y}{|E_x|} = \frac{800\text{ V/m}}{600\text{ V/m}} = \frac{4}{3} \Rightarrow \theta = 53.1^\circ \text{ above the } -x\text{-axis}$$

The electric field points $180^\circ - 53^\circ = 127^\circ$ counterclockwise (ccw) from the $+x$-axis.

30.55. Model: Assume the battery is ideal.

Visualize:

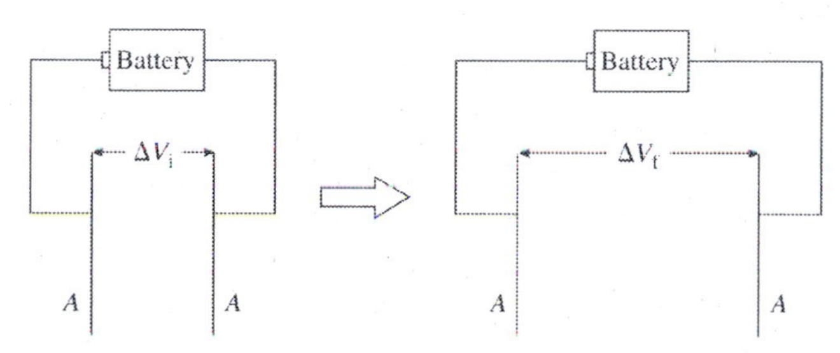

The pictorial representation shows the capacitor plates connected to a battery, and the capacitor plates moved apart with insulating handles while they are connected to the battery.

Solve: **(a)** The initial capacitance of the plates is

$$C_i = \frac{\varepsilon_0 A}{d_i} = \frac{\left(8.85\times10^{-12}\text{ C}^2/\text{N m}^2\right)\left(0.020\text{ m}\right)^2}{1.0\times10^{-3}\text{ m}} = 3.54\times10^{-12}\text{ F}$$

Consequently, an initial voltage $\Delta V_i = 9.0$ V charges the plates to

$$Q = \pm C_i\Delta V_i = \pm\left(3.54\times10^{-12}\text{ F}\right)\left(9.0\text{ V}\right) = \pm31.9\times10^{-12}\text{ C} = \pm32\text{ pC}$$

(b) The new capacitance is $C_f = \varepsilon_0 A/2d_i = \frac{1}{2}C_i$. The potential difference across the plates is determined by the battery and is unchanged: $\Delta V_f = \Delta V_i = 9.0$. Thus, the new charge on the plates is

$$Q = \pm C_i\Delta V_f = \pm\tfrac{1}{2}\left(3.54\times10^{-12}\text{ F}\right)\left(9.0\text{ V}\right) = \pm16.0\times10^{-12}\text{ C} = \pm16.0\text{ pC}$$

30.59. Visualize:

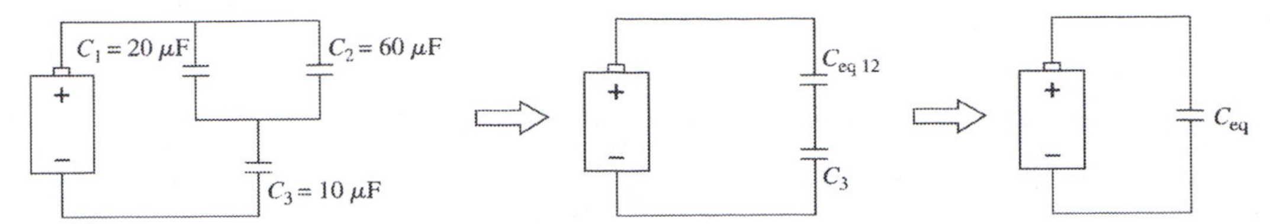

The pictorial representation shows how to find the equivalent capacitance of the three capacitors shown in the figure.
Solve: Because C_1 and C_2 are in parallel, their equivalent capacitance $C_{eq\ 12}$ is

$$C_{eq\ 12} = C_1 + C_2 = 20\ \mu F + 60\ \mu F = 80\ \mu F$$

Then, $C_{eq\ 12}$ and C_3 are in series. So,

$$\frac{1}{C_{eq}} = \frac{1}{C_{eq\ 12}} + \frac{1}{C_3} = \frac{1}{80\ \mu F} + \frac{1}{10\ \mu F} = \frac{9}{80}(\mu F)^{-1} \Rightarrow C_{eq} = \frac{80}{9}\ \mu F\ =\ 8.9\ \mu F$$

30.61. Model: Assume that the battery is ideal.
Visualize:

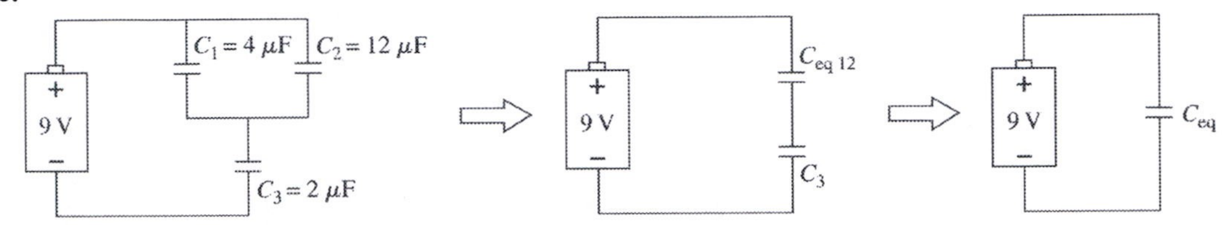

The pictorial representation shows how to find the equivalent capacitance of the three capacitors shown in the figure.
Solve: Because C_1 and C_2 are in parallel,

$$C_{eq\ 12} = C_1 + C_2 = 4\ \mu F + 12\ \mu F = 16\ \mu F$$

$C_{eq\ 12}$ and C_3 are in series, so

$$\frac{1}{C_{eq}} = \frac{1}{C_{eq\ 12}} + \frac{1}{C_3} = \frac{1}{16\ \mu F} + \frac{1}{2\ \mu F} = \frac{18}{32}(\mu F)^{-1} \Rightarrow C_{eq} = \tfrac{32}{18}\ \mu F$$

A potential difference of $\Delta V_C = 9$ V across a capacitor of equivalent capacitance $\tfrac{32}{18}\ \mu F$ produces a charge

$$Q = C_{eq}\Delta V_C = (\tfrac{32}{18}\ \mu F)9\ V\ =\ 16\ \mu C$$

Because C_{eq} is a series combination of two capacitors $C_{eq\ 12}$ and C_3, $Q_3 = Q_{eq\ 12} = 16\ \mu C$. The potential difference across C_3 is

$$\Delta V_3 = \frac{Q_3}{C_3} = \frac{16\ \mu C}{2\ \mu F} = 8.0\ V$$

Now, $Q_{eq\ 12} = 16\ \mu C$ is the charge on the equivalent capacitor with $C_{eq\ 12} = 16\ \mu F$. So, the potential difference across the equivalent capacitor $C_{eq\ 12}$ is

$$\Delta V_{eq\ 12} = \frac{Q_{eq\ 12}}{C_{eq\ 12}} = \frac{16\ \mu C}{16\ \mu F} = 1.0\ V$$

Parallel capacitors C_1 and C_2 have the same potential difference as the equivalent capacitor $C_{eq\ 12}$, so $\Delta V_1 = \Delta V_2 =$ 1.0 V. The charge on each is given by $Q = C\Delta V$, so $Q_1 = (4\ \mu F)(1.0\ V) = 4.0\ \mu C$ and $Q_2 = (12\ \mu F)(1.0\ V) = 12.0\ \mu C$. In summary, $Q_1 = 4.0\ \mu C$, $\Delta V_1 = 1.0$ V; $Q_2 = 12.0\ \mu C$, $\Delta V_2 = 1.0$ V; and $Q_3 = 16.0\ \mu C$, $\Delta V_3 = 8.0$ V.
Assess: Note that $\Delta V_3 + \Delta V_{eq\ 12} = 9.0\ V = \Delta V_{bat}$, as it should. Also that $Q_1 + Q_2 = 16.0\ \mu C = Q_{eq\ 12}$, as it should.

30.65. Model: Assume the battery is ideal.
Visualize:

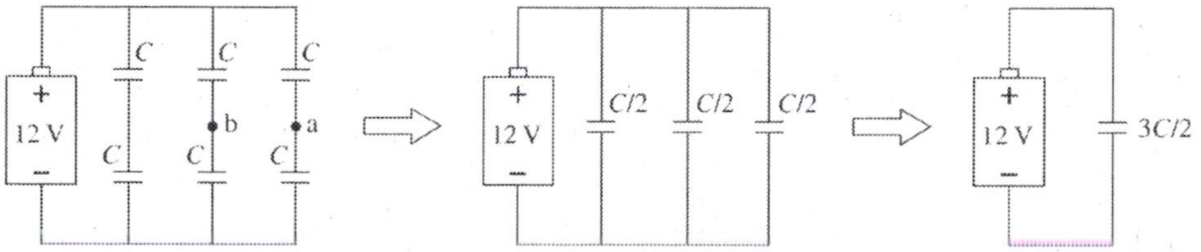

The circuit in Figure P30.65 has been redrawn to show that the six capacitors are arranged in three parallel combinations, each combination being a series combination of two capacitors.

Solve: **(a)** The equivalent capacitance of the two capacitors in series is $\frac{1}{2}C$. The equivalent capacitance of the six capacitors is $\frac{3}{2}C$.

(b) As points a and b are midpoints of identical capacitors, $V_a = V_b = 6.0$ V. Therefore, the potential difference between points a and b is zero.

30.71. **Solve:** **(a)** The capacitance of the parallel-plate capacitor is

$$C_1 = \frac{A\varepsilon_0}{d_1} = \frac{(0.10\text{ m}\times 0.10\text{ m})(8.85\times10^{-12}\text{ C}^2/\text{N m}^2)}{1.0\times10^{-3}\text{ m}} = 8.85\times10^{-11}\text{ F}$$

The electric potential energy stored in the capacitor is

$$U_1 = \frac{1}{2}\frac{Q^2}{C_1} = \frac{1}{2}\frac{(10\times10^{-9}\text{ C})^2}{8.85\times10^{-11}\text{F}} = 5.7\times10^{-7}\text{ J}$$

(b) There is no change in the charge. The energy change is due to the change in the capacitance. The new capacitance is

$$C_2 = \frac{\varepsilon_0 A}{d_2} = \frac{\varepsilon_0 A}{2d_1} = \frac{C_1}{2}$$

The amount of energy stored is $U_2 = 11.4\times10^{-7}$ J.

(c) Work was done on the capacitor by the agent pulling the plates apart, thereby adding energy into the system.

30.73. **Model:** Conservation of energy.

Solve: The energy stored in the capacitor is dissipated through the flash lamp. Power is the rate at which energy is dissipated/absorbed, so the energy dissipated by the flash lamp is

$$P\Delta t = (10\text{ W})(10\ \mu\text{s}) = 1.0\times10^{-4}\text{ J}$$

This is the electric potential energy in the capacitor. Using $U_C = \frac{1}{2}C\Delta V_C^2$,

$$1.0\times10^{-4}\text{ J} = \tfrac{1}{2}C(3.0\text{ V})^2 \Rightarrow C = 22\ \mu\text{F}$$

30.79. **Solve:** **(a)** A capacitor is constructed with two 10 cm × 10 cm plates. When the capacitor is connected to a 100 V source, ±400 nC is observed on the plates. What is the separation of the plates?

(b) From the first equation,

$$C = \frac{400\times10^{-9}\text{ C}}{100\text{ V}} = 4.0\times10^{-9}\text{ F}$$

Thus, the second equation becomes

$$4.0\times10^{-9}\text{ F} = \frac{(8.85\times10^{-12}\text{ C}^2/\text{N m}^2)(0.10\text{ m}\times0.10\text{ m})}{d} \Rightarrow d = 2.2\times10^{-5}\text{ m} = 0.022\text{ mm}$$

CURRENT AND RESISTANCE

31

Exercises and Problems

31.3. Solve: Equation 31.2 is $N_e = nAv_d\Delta t$. Using Table 31.1 for the electron density, we get

$$A = \frac{\pi D^2}{4} = \frac{N_e}{nv_d\Delta t}$$

$$\Rightarrow D = \sqrt{\frac{4\,N_e}{\pi n v_d \Delta t}} = \sqrt{\frac{4\left(1.0\times10^{16}\right)}{\pi\left(5.8\times10^{28}\text{ m}^{-3}\right)\left(8.0\times10^{-4}\text{ m/s}\right)\left(320\times10^{-6}\text{ s}\right)}} = 9.3\times10^{-4}\text{ m} = 0.93\text{ mm}$$

31.7. Solve: For $L = 1.0\text{ cm} = 1.0\times10^{-2}\text{ m}$, the surface area of the wire is

$$A = (2\pi r)L = \pi DL = \pi(1.0\times10^{-3}\text{ m})(1.0\times10^{-2}\text{ m}) = (1.0\times10^{-5}\text{ m}^2)\pi$$

The surface charge density of the wire is

$$\eta = \frac{Q}{A} = \frac{\left(1000\text{ cm}^{-1}\times1\text{ cm}\right)1.60\times10^{-19}\text{ C}}{\left(1.0\times10^{-5}\text{ m}^2\right)\pi} = 5.1\times10^{-12}\text{ C/m}^2$$

31.9. Model: We will use the model of conduction to relate the electric field strength to the mean free time between collisions.
Solve: From Equation 31.8, the electric field is

$$E = \frac{mi}{ne\tau A} = \frac{\left(9.11\times10^{-31}\text{ kg}\right)\left(5.0\times10^{19}\text{ s}^{-1}\right)}{\left(8.5\times10^{28}\text{ m}^{-3}\right)\left(1.6\times10^{-19}\text{ C}\right)\left(4.2\times10^{-15}\text{ s}\right)\pi\left(0.9\times10^{-3}\text{ m}\right)^2} = 0.31\text{ N/C}$$

31.13. Solve: **(a)** From Equation 31.13 and Table 31.1, the current density in the gold wire is

$$J = nev_d = \left(5.9\times10^{28}\text{ m}^{-3}\right)\left(1.60\times10^{-19}\text{ C}\right)\left(3.0\times10^{-4}\text{ m/s}\right) = 2.83\times10^{6}\text{ A/m}^2$$

The current density is $2.8\times10^6\text{ A/m}^2$.
(b) The current is

$$I = JA = \left(2.83\times10^{6}\text{ A/m}^2\right)\pi\left(0.25\times10^{-3}\text{ m}\right)^2 = 0.56\text{ A}$$

31.19. Visualize:

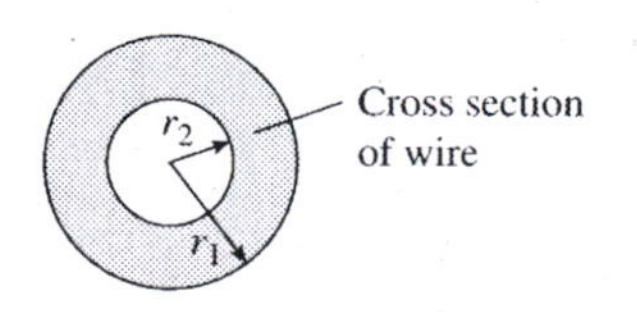

Solve: The current-carrying cross section of the wire is

$$A = \pi r_1^2 - \pi r_2^2 = \pi\left[(0.0010\text{ m})^2 - (0.00050\text{ m})^2\right] = 2.356\times10^{-6}\text{ m}^2$$

The current density is

$$J = \frac{10\text{ A}}{2.356\times10^{-6}\text{ m}^2} = 4.2\times10^6\text{ A/m}^2$$

31.21. Model: We will use the model of conduction to relate the mean time between collisions to conductivity.
Solve: From Equation 31.17, Table 31.1, and Table 31.2, the mean time between collisions for silver is

$$\tau_{\text{silver}} = \frac{m\sigma_{\text{silver}}}{n_{\text{silver}}e^2} = \frac{(9.11\times10^{-31}\text{ kg})(6.2\times10^7\ \Omega^{-1}\text{m}^{-1})}{(5.8\times10^{28}\text{ m}^{-3})(1.60\times10^{-19}\text{ C})^2} = 3.8\times10^{-14}\text{ s}$$

Similarly, the mean time between collisions for gold is $\tau_{\text{gold}} = 2.5\times10^{-14}$ s.

Assess: Mean free times in metals are of the order of $\approx10^{-14}$ s.

31.25. Solve: From Equations 31.18 and 31.19, the resistivity is

$$\rho = \frac{E}{J} = \frac{E}{I/A} = \frac{EA}{I} = \frac{E\pi r^2}{I} = \frac{(0.0075\text{ V/m})\pi(0.50\times10^{-3}\text{ m})^2}{3.9\times10^{-3}\text{ A}} = 1.51\times10^{-6}\ \Omega\text{ m}$$

From Table 31.2, we see that the wire is made of nichrome.

31.31. Solve: **(a)** Using Table 31.2 and Equation 31.23, the resistance is

$$R = \frac{\rho L}{A} = \frac{\rho L}{\pi r^2} = \frac{(1.7\times10^{-8}\ \Omega\text{m})(1.0\text{ m})}{\pi(2.5\times10^{-4}\text{ m})^2} = 0.087\ \Omega$$

(b) The resistance is

$$R = \frac{\rho L}{A} = \frac{\rho L}{d^2} = \frac{(3.5\times10^{-5}\ \Omega\text{m})(0.10\text{ m})}{(0.0010\text{ m})^2} = 3.5\ \Omega$$

31.35. Model: Assume the battery is an ideal battery.
Solve: We can find the current I from Equation 31.24 provided we know the resistance R of the gold wire. From Equation 31.23 and Table 31.2, the resistance of the wire is

$$R = \frac{\rho L}{A} = \frac{\rho L}{\pi r^2} = \frac{(2.4\times10^{-8}\ \Omega\text{m})(100\text{ m})}{\pi(0.050\times10^{-3}\text{ m})^2} = 305.6\ \Omega \Rightarrow I = \frac{\Delta V_{\text{wire}}}{R} = \frac{0.70\text{ V}}{305.6\ \Omega} = 2.3\text{ mA}$$

31.39. Solve: We need an aluminum wire whose resistance and length are the same as that of a 0.50-mm-diameter copper wire. That is,

$$R_{\text{Cu}} = \frac{\rho_{\text{Cu}}L}{A_{\text{Cu}}} = R_{\text{Al}} = \frac{\rho_{\text{Al}}L}{A_{\text{Al}}} \Rightarrow A_{\text{Al}} = \left(\frac{\rho_{\text{Al}}}{\rho_{\text{Cu}}}\right)A_{\text{Cu}} \Rightarrow \pi r_{\text{Al}}^2 = \left(\frac{\rho_{\text{Al}}}{\rho_{\text{Cu}}}\right)\pi r_{\text{Cu}}^2$$

$$\Rightarrow r_{\text{Al}} = \sqrt{\frac{\rho_{\text{Al}}}{\rho_{\text{Cu}}}}r_{\text{Cu}} = \sqrt{\frac{2.8\times10^{-8}\ \Omega\text{ m}}{1.7\times10^{-8}\ \Omega\text{ m}}}(0.25\text{ mm}) = 0.32\text{ mm}$$

We need a 0.64-mm-diameter aluminum wire.

31.43. Visualize: Please refer to Figure P31.43.
Solve: **(a)** The current associated with the moving film is the rate at which the charge on the film moves past a certain point. The tangential speed of the film is

$$v = \omega r = (90\text{ rpm})(4.0\text{ cm}) = 90\frac{\text{rev}}{\text{min}}\times\frac{1\text{ min}}{60\text{ s}}\times\frac{2\pi\text{ rad}}{1\text{ rev}}\times1.0\text{ cm} = 9.425\text{ cm/s}$$

In 1.0 s the film moves a distance of 9.425 cm. This means the area of the film that moves to the right in 1.0 s is (9.425 cm)(4.0 cm) = 37.7 cm^2. The amount of charge that passes to the right in 1.0 s is

$$Q = (37.7\ \text{cm}^2)(-2.0 \times 10^{-9}\ \text{C/cm}^2) = -75.4 \times 10^{-9}\ \text{C}$$

Since $I = Q/\Delta t$, we have

$$I = \frac{\left|-\left(75.4\times10^{-9}\ \text{C}\right)\right|}{1\ \text{s}} = 75.4\ \text{nA}$$

The current is 75 nA.

(b) Having found the current in part (a), we can once again use $I = Q/\Delta t$ to obtain Δt:

$$\Delta t = \frac{Q}{I} = \frac{\left|-10\times10^{-6}\ \text{C}\right|}{75.4\times10^{-9}\ \text{A}} = 133\ \text{s}$$

31.45. Model: The current is the rate at which the charge of the ions moves through the ionic solution.
Solve: Because the atomic mass of gold is 197 g, the number of gold atoms is

$$N = \frac{M}{M_{\text{A}}}N_{\text{A}} = \left(\frac{0.50\ \text{g}}{197\ \text{g mol}^{-1}}\right)6.02\times10^{23}\ \text{mol}^{-1} = 1.53\times10^{21}\ \text{atoms}$$

We need to deposit $N = 1.53 \times 10^{21}$ gold ions, each with a charge of -1.60×10^{-19} C, in 3.0 hours on the statue. The current is

$$I = \frac{Q}{\Delta t} = \frac{\left(1.53\times10^{21}\right)\left(1.60\times10^{-19}\ \text{C}\right)}{3.0\times3600\ \text{s}} = 23\ \text{mA}$$

31.49. Model: We assume that the charge carriers are uniformly distributed throughout the wire.
Solve: Using Equation 31.16, we can write the current density as

$$J = \frac{I}{A} = \frac{ne^2\tau}{m}E \Rightarrow I = \left(\frac{ne^2\tau}{m}E\right)A$$

For a given wire, the current is thus proportional to the area through which the current flows: $I_{\text{total}} \propto R^2$ and $I_{\text{center}} \propto \left(\frac{1}{2}R\right)^2$. Therefore,

$$\frac{I_{\text{center}}}{I_{\text{total}}} = \frac{1}{4} = 25\%$$

That is, 25% of the total current flows in the part of the wire with radius $r \le R/2$.

31.51. Visualize:

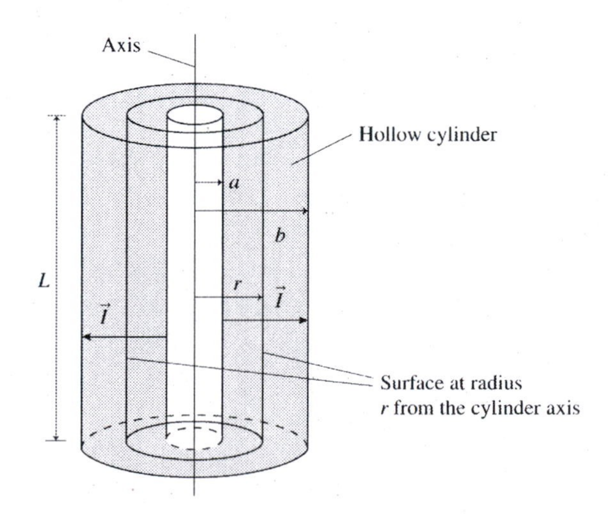

Solve: **(a)** Consider a cylindrical surface inside the metal at a radial distance r from the center. The current is flowing through the walls of this cylinder, which have surface area $A=(2\pi r)L$. Thus

$$I = JA = \sigma E(2\pi r)L$$

Thus the electric field strength at radius r is

$$E = \frac{I}{2\pi\sigma Lr}$$

(b) For iron, with $\sigma = 1.0\times10^{7}\ \Omega^{-1}\text{m}^{-1}$,

$$E_{\text{inner}} = \left(\frac{25\text{ A}}{2\pi(0.10\text{ m})(1.0\times10^{7}\ \Omega^{-1}\text{m}^{-1})}\right)\left(\frac{1}{0.010\text{ m}}\right) = 4.0\times10^{-4}\text{ V/m}$$

$$E_{\text{outer}} = \left(\frac{25\text{ A}}{2\pi(0.10\text{ m})(1.0\times10^{7}\ \Omega^{-1}\text{m}^{-1})}\right)\left(\frac{1}{0.025\text{ m}}\right) = 1.59\times10^{-4}\text{ V/m}$$

31.57. Model: Because current is conserved, the current flowing in the 2.0-mm-diameter segment of the wire is the same as in the 1.0-mm-diameter segment.
Visualize: Please refer to Figure P31.57. We will denote all quantities for the 1.0-mm-diameter wire with the subscript 1, and all quantities for the 2.0-mm-diameter wire with the subscript 2.
Solve: Equation 31.13 is $J = nev_d$. This means the current densities in the two segments are

$$J_1 = nev_{d1} \qquad J_2 = nev_{d2}$$

Dividing these equations, we get $v_{d2} = (J_2/J_1)v_{d1}$. Because current is conserved, $I_1 = I_2 = 2.0$ A. So,

$$\frac{J_2}{J_1} = \frac{I_2/A_2}{I_1/A_1} = \frac{A_1}{A_2} \Rightarrow v_{d2} = \frac{A_1}{A_2}v_{d1} = \left(\frac{D_1}{D_2}\right)^2 v_{d1} = \left(\frac{1.0\text{ mm}}{2.0\text{ mm}}\right)^2(2.0\times10^{-4}\text{ m/s}) = 5.0\times10^{-5}\text{ m/s}$$

Assess: A drift velocity which is small and only $\left(\frac{1}{4}\right)$ of the drift velocity in the 1.0-mm-diameter wire is reasonable.

31.63. Model: Assume the battery is ideal.
Visualize: The current supplied by the battery and passing through the wire is $I = \Delta V_{\text{bat}}/R$. A graph of current versus time has exactly the same shape as the graph of ΔV_{bat} with an initial value of $I_0 = (\Delta V_{\text{bat}})_0/R = (1.5\text{ V})/(3.0\ \Omega) = 0.50$ A. The horizontal axis has been changed to seconds.

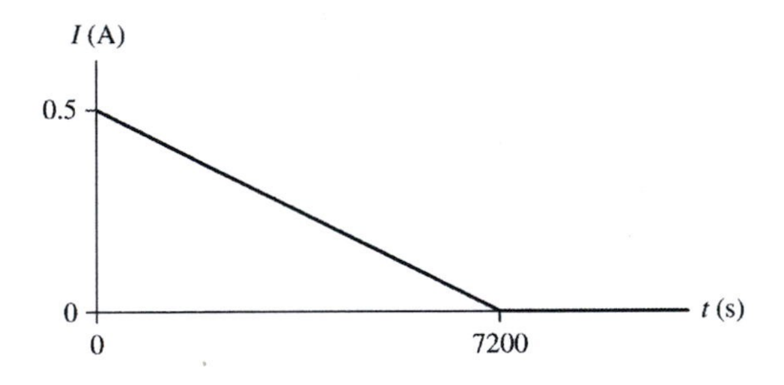

Solve: Current is $I = dQ/dt$. Thus the total charge supplied by the battery is

$$Q = \int_0^\infty I\,dt = \text{area under the current-versus-time graph}$$

$$= \tfrac{1}{2}(7200\text{ s})(0.50\text{ A}) = 1.80\times10^{3}\text{ C}$$

31.65. Model: The volume of the wire remains constant as it is stretched. The cross-sectional area of the wire changes uniformly as it stretches.

Solve: The resistance of the wire before it is stretched is

$$R = \frac{\rho L}{A} = \frac{\rho L^2}{AL} = \frac{\rho L^2}{V}.$$

The volume V remains constant as the wire is stretched. After stretching, the resistance is

$$R' = \frac{\rho L'^2}{V}.$$

Taking the ratio of these two equations and using the fact that ρ is a property of the material and therefore does not change,

$$\frac{R}{R'} = \frac{L^2}{L'^2} = \frac{L^2}{(2L)^2} = \frac{1}{4} \Rightarrow R' = 4R$$

The wire's resistance is $4R$.

Assess: Stretching a wire increases the one dimension of length but decreases the two dimensions of cross-sectional area, so the resistance increases.

Fundamentals of Circuits

32

Exercises and Problems

32.1. Solve:

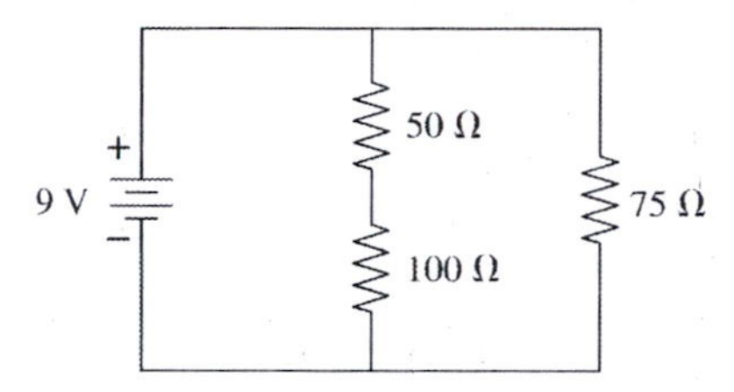

From the circuit in Figure EX32.1, we see that 50 Ω and 100 Ω resistors are connected in series across the battery. Another resistor of 75 Ω is also connected across the battery.

32.5. Model: Assume ideal connecting wires and an ideal battery for which $\Delta V_{bat} = \mathcal{E}$.

Visualize: Please refer to Figure EX32.5. We will choose a clockwise direction for I. Note that the choice of the current's direction is arbitrary because, with two batteries, we may not be sure of the actual current direction. The 3 V battery will be labeled 1 and the 6 V battery will be labeled 2.

Solve: **(a)** Kirchhoff's loop law, going clockwise from the negative terminal of the 3-V battery is

$$\Delta V_{\text{closed loop}} = \sum_i (\Delta V)_i = \Delta V_{\text{bat 1}} + \Delta V_{\text{R}} + \Delta V_{\text{bat 2}} = 0$$

$$\Rightarrow +3\ \text{V} - (18\ \Omega)\,I + 6\ \text{V} = 0 \Rightarrow I = \frac{9\ \text{V}}{18\ \Omega} = 0.5\ \text{A}$$

Thus, the current through the 18 Ω resistor is 0.5 A. Because I is positive, the current is left to right (i.e., clockwise).

(b)

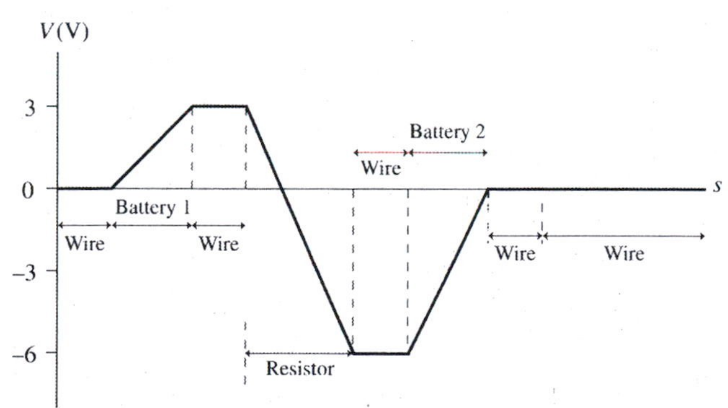

Assess: The graph shows a 3 V gain in battery 1, a –9 V loss in the resistor, and a gain of 6 V in battery 2. The final potential is the same as the initial potential, as required.

32.9. Model: The 100 W rating is for operating at 120 V.
Solve: A standard bulb uses $\Delta V = 120$ V. We can use the power dissipation to find the resistance of the filament:

$$P = \frac{\Delta V^2}{R} \Rightarrow R = \frac{\Delta V^2}{P} = \frac{(120\text{ V})^2}{100\text{ W}} = 144\ \Omega$$

But the resistance is related to the filament's geometry:

$$R = \frac{\rho L}{A} = \frac{\rho L}{\pi r^2} \Rightarrow r = \sqrt{\frac{\rho L}{\pi R}} = \sqrt{\frac{(9.0\times10^{-7}\ \Omega\text{ m})(0.070\text{ m})}{\pi(144\ \Omega)}} = 1.18\times10^{-5}\text{ m} = 11.8\ \mu\text{m}$$

The filament's diameter is $d = 2r = 23.6\ \mu$m.

32.11. Solve: (a) The average power consumed by a typical American family is

$$P_{\text{avg}} = 1000\frac{\text{kWh}}{\text{month}} = 1000\frac{\text{kWh}}{30\times24\text{ h}} = \frac{1000}{720}\text{ kW} = 1.389\text{ kW}$$

Because $P = (\Delta V)I$ with ΔV as the voltage of the power line to the house,

$$I_{\text{avg}} = \frac{P_{\text{avg}}}{\Delta V} = \frac{1.389\text{ kW}}{120\text{ V}} = \frac{1389\text{ W}}{120\text{ V}} = 11.6\text{ A}$$

(b) Because $P = (\Delta V)^2/R$,

$$R_{\text{avg}} = \frac{(\Delta V)^2}{P_{\text{avg}}} = \frac{(120\text{ V})^2}{1389\text{ W}} = 10.4\ \Omega$$

32.15. Model: Assume ideal connecting wires and an ideal power supply.
Visualize:

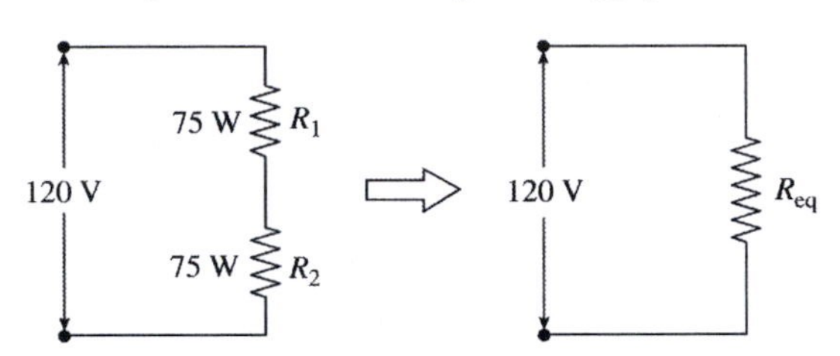

The two light bulbs are basically two resistors in series.
Solve: A 75 W (120 V) light bulb has a resistance of

$$R = \frac{\Delta V^2}{P} = \frac{(120\text{ V})^2}{75\text{ W}} = 192\ \Omega$$

The combined resistance of the two bulbs is

$$R_{\text{eq}} = R_1 + R_2 = 192\ \Omega + 192\ \Omega = 384\ \Omega$$

The current I flowing through R_{eq} is

$$I = \frac{\Delta V}{R_{\text{eq}}} = \frac{120\text{ V}}{384\ \Omega} = 0.3125\text{ A}$$

Because R_{eq} is a series combination of R_1 and R_2, the current 0.3125 A flows through R_1 and R_2. Thus,

$$P_{\text{R1}} = I^2R_1 = (0.3125\text{ A})^2(192\ \Omega) = 18.8\text{ W} = P_{\text{R2}}$$

32.17. Model: Assume ideal connecting wires and an ideal ammeter but not an ideal battery.
Visualize: Please refer to Figure EX32.17.
Solve: An ideal ammeter has zero resistance, so the battery is being short circuited. If I is the current flowing through the circuit, then

$$I = \frac{\mathcal{E}}{r} \Rightarrow r = \frac{\mathcal{E}}{I} = \frac{1.5\text{ V}}{2.3\text{ A}} = 0.65\ \Omega$$

The power dissipated by the internal resistance is

$$P = I^2 r = (2.3\text{ A})^2(0.6522\ \Omega) = 3.5\text{ W}$$

32.21. Visualize: The three resistors in Figure EX32.21 are equivalent to a resistor of resistance $R_{eq} = 75\ \Omega$.

Solve: Because the three resistors are in parallel,

$$\frac{1}{R_{eq}} = \frac{1}{R} + \frac{1}{200\ \Omega} + \frac{1}{R} = \frac{2}{R} + \frac{1}{200\ \Omega} = \frac{400\ \Omega + R}{(200\ \Omega)R} \Rightarrow R_{eq} = 75\ \Omega = \frac{(200\ \Omega)R}{(400\ \Omega + R)} = \frac{200\ \Omega}{1 + \left(\frac{400\ \Omega}{R}\right)}$$

$$\Rightarrow R = \frac{400\ \Omega}{\frac{200\ \Omega}{75\ \Omega} - 1} = 240\ \Omega$$

32.23. Model: The connecting wires are ideal with zero resistance.

Solve:

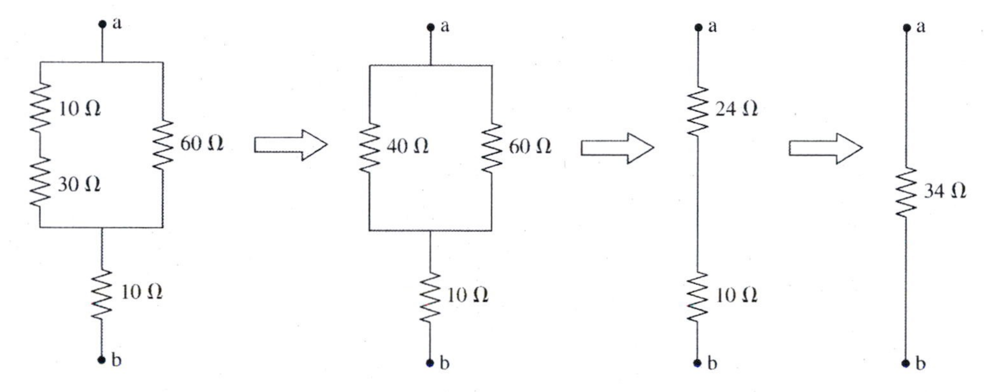

For the first step, the 10 Ω and 30 Ω resistors are in series and the equivalent resistance is 40 Ω. For the second step, the 60 Ω and 40 Ω resistors are in parallel and the equivalent resistance is

$$\left[\frac{1}{40\ \Omega} + \frac{1}{60\ \Omega}\right]^{-1} = 24\ \Omega$$

For the third step, the 24 Ω and 10 Ω resistors are in series and the equivalent resistance is 34 Ω.

32.25. Model: The connecting wires are ideal with zero resistance.

Solve:

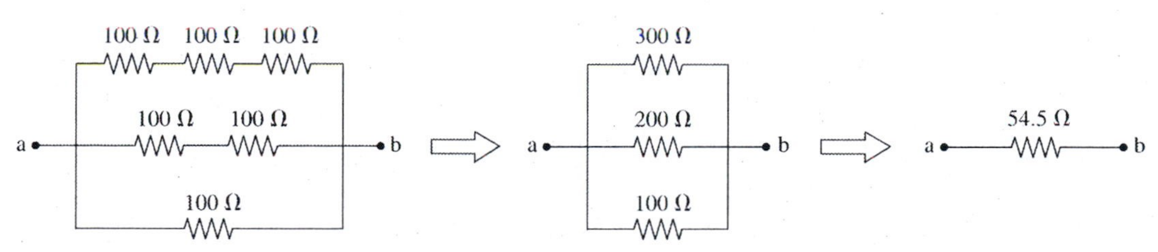

In the first step, the resistors 100 Ω, 100 Ω, and 100 Ω in the top branch are in series. Their combined resistance is 300 Ω. In the middle branch, the two resistors, each 100 Ω, are in series. So, their equivalent resistance is 200 Ω. In the second step, the three resistors are in parallel. Their equivalent resistance is

$$\frac{1}{R_{eq}} = \frac{1}{300\ \Omega} + \frac{1}{200\ \Omega} + \frac{1}{100\ \Omega} \Rightarrow R_{eq} = 54.5\ \Omega$$

The equivalent resistance of the circuit is 54.5 Ω.

32.27. Model: Grounding does not affect a circuit's behavior.

Visualize: Please refer to Figure EX32.27.

Solve: Because the earth has $V_{\text{earth}} = 0$ V, point d has a potential of zero. In going from point d to point a, the potential increases by 9 V. Thus, point a is at a potential of 9 V. Let us calculate the current I in the circuit before calculating the potentials at points b and c. Applying Kirchhoff's loop rule, starting clockwise from point d,

$$\sum_i (\Delta V)_i = \Delta V_{9\text{ V bat}} + \Delta V_{2\,\Omega} + \Delta V_{6\text{ V bat}} + \Delta V_{1\,\Omega} = 0$$

$$\Rightarrow +9\text{ V} - I(2\,\Omega) - 6\text{ V} - I(1\,\Omega) = 0 \Rightarrow I = \frac{3\text{ V}}{3\,\Omega} = 1\text{ A}$$

There is a drop in potential from point a to point b by an amount $IR = (1\text{ A})(2\,\Omega) = 2$ V. Thus, the potential at point b is 9 V − 2 V = 7 V. The potential decreases from 7 V at point b to 7 V − 6 V = 1 V at point c. There is a further decrease in potential across the 1 Ω resistor of $IR = (1\text{ A})(1\,\Omega) = 1$ V. That is, the potential of 1 V at c becomes 0 V at point d, as it must. In summary, the potentials at a, b, c, and d are 9 V, 7 V, 1 V, and 0 V.

32.31. Model: Assume ideal wires as the capacitors discharge through the two 1 kΩ resistors.
Visualize: The circuit in Figure EX32.31 has an equivalent circuit with resistance R_{eq} and capacitance C_{eq}.
Solve: The equivalent capacitance is $C_{\text{eq}} = 2\ \mu\text{F} + 2\ \mu\text{F} = 4\ \mu\text{F}$, and the equivalent resistance is

$$\frac{1}{R_{\text{eq}}} = \frac{1}{1\text{ k}\Omega} + \frac{1}{1\text{ k}\Omega} \Rightarrow R_{\text{eq}} = 0.5\text{ k}\Omega$$

Thus, the time constant for the discharge of the capacitors is

$$\tau = R_{\text{eq}}C_{\text{eq}} = (0.5\text{ k}\Omega)(4\ \mu\text{F}) = 2 \times 10^{-3}\text{ s} = 2\text{ ms}$$

32.35. Model: A capacitor discharges through a resistor. Assume ideal wires.
Solve: The discharge current or the resistor current follows Equation 32.35: $I = I_0 e^{-t/RC}$. We wish to find the capacitance C so that the resistor current will decrease to 25% of its initial value in 2.5 ms. That is,

$$0.25\,I_0 = I_0 e^{-2.5\text{ ms}/(100\,\Omega)C} \Rightarrow \ln(0.25) = -\frac{2.5\times10^{-3}\text{ s}}{(100\,\Omega)C} \Rightarrow C = 18.0\ \mu\text{F}$$

32.41. Visualize:

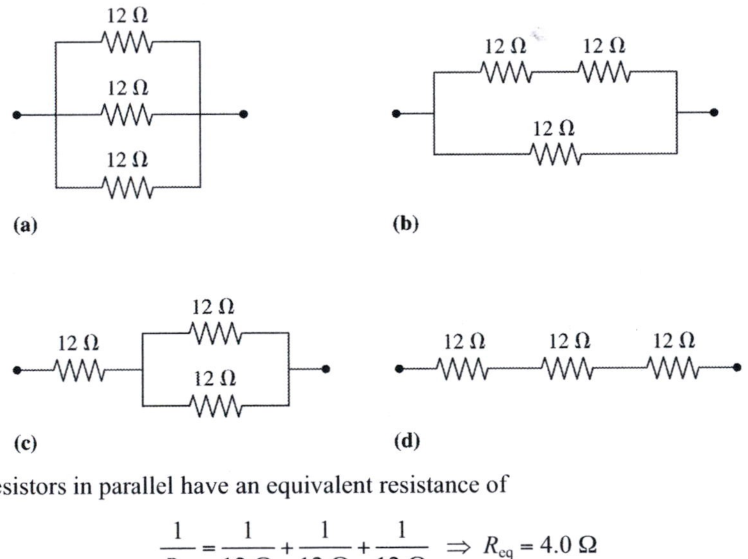

12 Ω, 12 Ω, 12 Ω

(a)

12 Ω, 12 Ω, 12 Ω

(b)

12 Ω, 12 Ω, 12 Ω

(c)

12 Ω, 12 Ω, 12 Ω

(d)

Solve: **(a)** The three resistors in parallel have an equivalent resistance of

$$\frac{1}{R_{\text{eq}}} = \frac{1}{12\,\Omega} + \frac{1}{12\,\Omega} + \frac{1}{12\,\Omega} \Rightarrow R_{\text{eq}} = 4.0\,\Omega$$

(b) One resistor in parallel with two series resistors has an equivalent resistance of

$$\frac{1}{R_{\text{eq}}} = \frac{1}{12\,\Omega + 12\,\Omega} + \frac{1}{12\,\Omega} = \frac{1}{24\,\Omega} + \frac{1}{12\,\Omega} = \frac{1}{8\,\Omega} \Rightarrow R_{\text{eq}} = 8.0\,\Omega$$

(c) One resistor in series with two parallel resistors has an equivalent resistance of

$$\frac{1}{R_{eq}} = 12\ \Omega + \left(\frac{1}{12\ \Omega} + \frac{1}{12\ \Omega}\right)^{-1} = 12\ \Omega + 6\ \Omega = 18.0\ \Omega$$

(d) The three resistors in series have an equivalent resistance of

$$12\ \Omega + 12\ \Omega + 12\ \Omega = 36\ \Omega$$

32.43. Model: Assume the batteries and the connecting wires are ideal.
Visualize: Please refer to Figure P32.43.
Solve: (a) The two batteries in this circuit are oriented to "oppose" each other. The curent will flow in the direction of the battery with the greater voltage. The direction of the current is counterclockwise because the 12 V battery is greater.
(b) There are no junctions, so the same current I flows through all circuit elements. Applying Kirchhoff's loop law in the *counterclockwise* direction and starting at the lower right corner,

$$\sum\Delta V_i = 12\text{ V} - I(12\ \Omega) - I(6\ \Omega) - 6\text{ V} - IR = 0$$

Note that the IR terms are all negative because we're applying the loop law in the direction of current flow, and the potential *decreases* as current flows through a resistor. We can easily solve to find the unknown resistance R:

$$6\text{ V} - I(18\ \Omega) - IR = 0 \Rightarrow R = \frac{6\text{ V} - (18\ \Omega)I}{I} = \frac{6\text{ V} - (18\ \Omega)(0.25\text{ A})}{0.25\text{ A}} = 6\ \Omega$$

(c) The power is $P = I^2R = (0.25\text{ A})^2(6\ \Omega) = 0.38$ W.
(d)

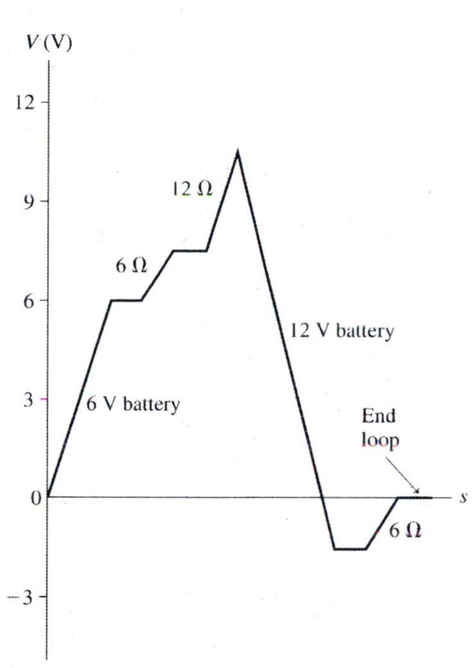

The potential difference across a resistor is $\Delta V = IR$, giving $\Delta V_6 = 1.5$ V, and $\Delta V_{12} = 3$ V. Starting from the lower left corner, the graph goes around the circuit *clockwise*, opposite from the direction in which we applied the loop law. In this direction, we speak of potential as *lost* in the batteries and *gained* in the resistors.

32.47. Model: Assume that the connecting wire and the battery are ideal.
Visualize: Please refer to Figure P32.47.
Solve: The middle and right branches are in parallel, so the potential difference across these two branches must be the same. The currents are known, so these potential differences are

$$\Delta V_{\text{middle}} = (3.0\text{ A})R = \Delta V_{\text{right}} = (2.0\text{ A})(R + 10\ \Omega)$$

This is easily solved to give $R = 20\ \Omega$. The middle resistor R is connected directly across the battery, thus (for an ideal battery, with no internal resistance) the potential difference ΔV_{middle} equals the emf of the battery. That is

$$\mathcal{E} = \Delta V_{\text{middle}} = (3.0\text{ A})(20\ \Omega) = 60\text{ V}$$

32.51. Model: Assume an ideal battery and ideal connecting wires.
Visualize:

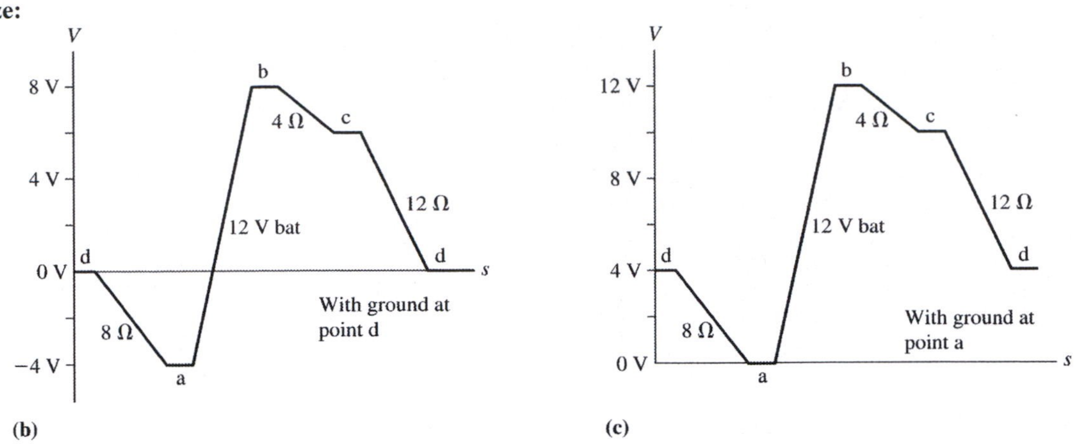

Solve: (a) Grounding one point doesn't affect the basic analysis of the circuit. In Figure P32.51, there is a single loop with a single current I flowing in the clockwise direction. Applying Kirchhoff's loop law clockwise from the lower right corner gives

$$\Sigma\Delta V_i = -8I + 12\text{ V} - 4I - 12I = 0\text{ V} \Rightarrow I = \frac{12\text{ V}}{24\ \Omega} = 0.50\text{ A}$$

Knowing the current, we can use $\Delta V = IR$ to find the potential difference across each resistor:

$$\Delta V_8 = 4\text{ V} \qquad \Delta V_4 = 2\text{ V} \qquad \Delta V_{12} = 6\text{ V}$$

The purpose of grounding one point in the circuit is to establish that point as the *specific* potential $V = 0$ V. Grounding point d makes that potential there $V_d = 0$ V. Then we can use the known potential differences to find *the* potential at other points in the circuit. Point a is 4 V *less* than point d (because potential decreases in the direction of current flow), so $V_a = V_d - 4\text{ V} = -4\text{ V}$. Point b is 12 V *more* than point a because of the battery. So $V_b = V_a + 12\text{ V} = 8\text{ V}$. Point c is 2 V *less* than point b, so $V_c = V_b - 2\text{ V} = 6\text{ V}$. Point d is 6 V *less* than point c, so $V_d = V_c - 6\text{ V} = 0\text{ V}$. This is a consistency check—making one complete loop brings us back to the potential at which we started, namely 0 V.
(b) The information about the potentials is shown in the graph above.
(c) Moving the ground to point a doesn't change the basic analysis of part (a) or the potential *differences* found there. All that changes is that now $V_a = 0$ V. Point b is 12 V *more* than point a because of the battery. So, $V_b = V_a + 12\text{ V} = 12\text{ V}$. Point c is 2 V *less* than point b, so $V_c = V_b - 2\text{ V} = 10\text{ V}$. Point d is 6 V *less* than point c, so $V_d = V_c - 6\text{ V} = 4\text{ V}$. Point a is 4 V *less* than point d, so $V_a = V_d - 4\text{ V} = 0\text{ V}$. This brings us back to where we started. The information about the potentials is shown in the graph above.

32.57. Model: The voltage source/battery and the connecting wires are ideal.
Visualize: Please refer to Figure P32.57.
Solve: Let us first apply Kirchhoff's loop law starting clockwise from the lower left corner:

$$+V_{in} - IR - I(100\ \Omega) = 0\text{ V} \Rightarrow I = \frac{V_{in}}{R + 100\ \Omega}$$

The output voltage is

$$V_{out} = (100\ \Omega)I = (100\ \Omega)\left(\frac{V_{in}}{R + 100\ \Omega}\right) \Rightarrow \frac{V_{out}}{V_{in}} = \frac{100\ \Omega}{R + 100\ \Omega}$$

For $V_{out} = V_{in}/10$, the above equation can be simplified to obtain R:

$$\frac{V_{in}/10}{V_{in}} = \frac{100\ \Omega}{R + 100\ \Omega} \Rightarrow R + 100\ \Omega = 1000\ \Omega \Rightarrow R = 900\ \Omega$$

32.61. Model: The battery and the connecting wires are ideal.

Visualize:

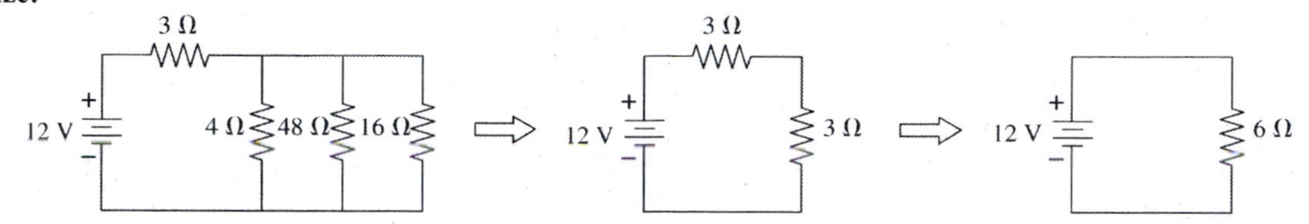

The figure shows how to simplify the circuit in Figure P32.61 using the laws of series and parallel resistances. Having reduced the circuit to a single equivalent resistance, we will reverse the procedure and "build up" the circuit using the loop law and the junction law to find the current and potential difference of each resistor.

Solve: From the last circuit in the diagram,

$$I = \frac{\mathcal{E}}{6\ \Omega} = \frac{12\ \text{V}}{6\ \Omega} = 2\ \text{A}$$

Thus, the current through the battery is 2 A. As we rebuild the circuit, we note that series resistors *must* have the same current I and that parallel resistors *must* have the same potential difference ΔV.

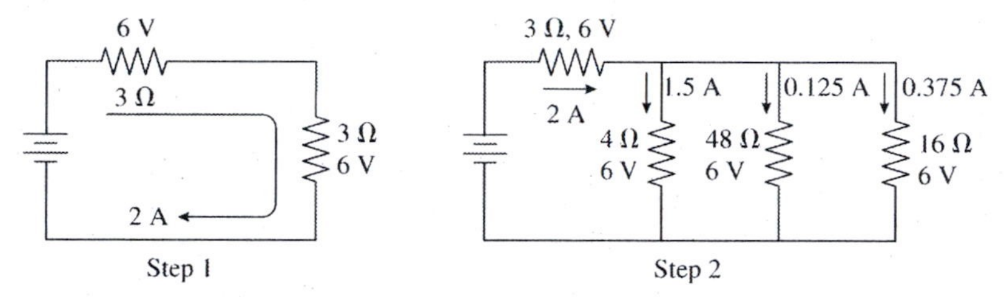

In Step 1, the 6 Ω resistor is returned to a 3 Ω and 3 Ω resistor in series. Both resistors must have the same 2 A current as the 6 Ω resistance. We then use Ohm's law to find

$$\Delta V_3 = (2\ \text{A})(3\ \Omega) = 6\ \text{V}$$

As a check, 6 V + 6 V = 12 V, which was ΔV of the 6 Ω resistor. In Step 2, one of the two 3 Ω resistances is returned to the 4 Ω, 48 Ω, and 16 Ω resistors in parallel. The three resistors must have the same $\Delta V = 6$ V. From Ohm's law,

$$I_4 = \frac{6\ \text{V}}{4\ \Omega} = 1.5\ \text{A} \qquad I_{48} = \frac{6\ \text{V}}{48\ \Omega} = \frac{1}{8}\ \text{A} \qquad I_{16} = \frac{6\ \text{V}}{16\ \Omega} = \frac{3}{8}\ \text{A}$$

Resistor	Potential difference (V)	Current (A)
3 Ω	6	2
4 Ω	6	1.5
48 Ω	6	1.2
16 Ω	6	3.8

32.63. Model: The batteries and the connecting wires are ideal.

Visualize:

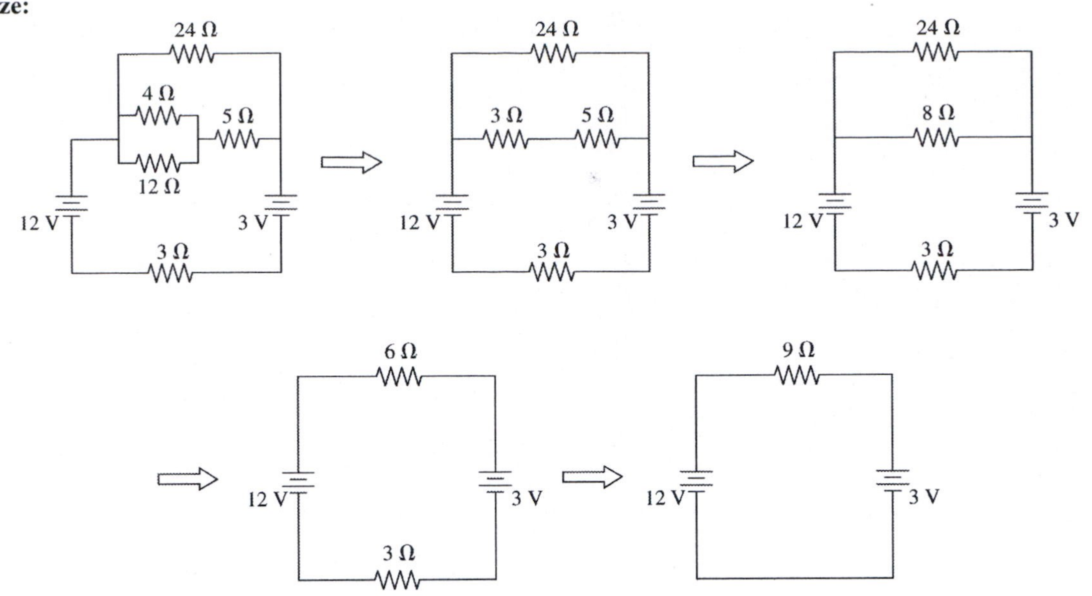

The figure shows how to simplify the circuit in Figure P32.63 using the laws of series and parallel resistances. Having reduced the circuit to a single equivalent resistance, we will reverse the procedure and "build up" the circuit using the loop law and the junction law to find the current and potential difference of each resistor.

Solve: From the last circuit in the figure and from Kirchhoff's loop law,

$$I = \frac{12\text{ V} - 3\text{ V}}{9\ \Omega} = 1.0\text{ A}$$

Thus, the current through the batteries is 1.0 A. As we rebuild the circuit, we note that series resistors *must* have the same current I and that parallel resistors *must* have the same potential difference.

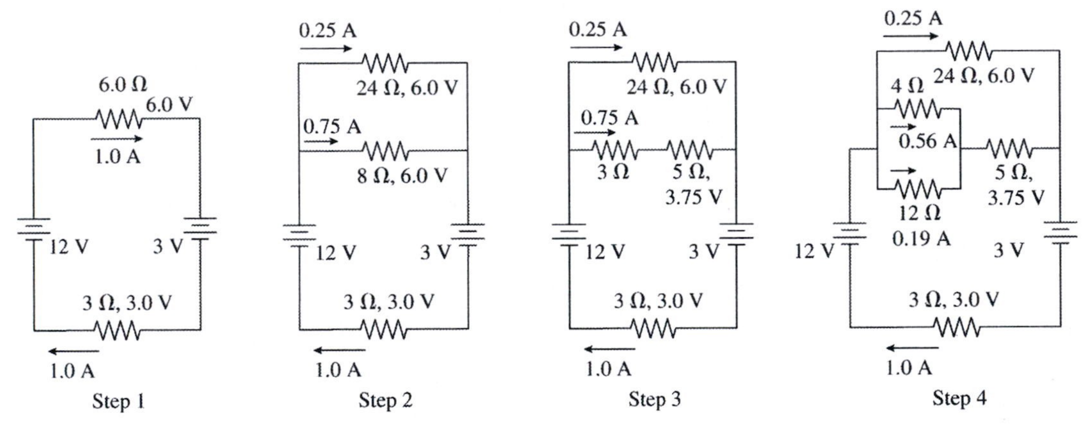

In Step 1 of the above figure, the $9\ \Omega$ resistor is returned to the $6\ \Omega$ and $3\ \Omega$ resistors in series. Both resistors must have the same 1.0 A current as the $6\ \Omega$ resistor. We use Ohm's law to find

$$\Delta V_3 = (1.0\text{ A})(3\ \Omega) = 3.0\text{ V} \qquad \Delta V_6 = 6.0\text{ V}$$

As a check, 3.0 V + 6.0 V = 9 V, which was $\Delta V = (12\text{ V} - 3\text{ V}) = 9$ V of the $9\ \Omega$ resistor. In Step 2, the $6\ \Omega$ resistor is returned to the $24\ \Omega$ and $8\ \Omega$ resistors in parallel. The two resistors must have the same potential difference $\Delta V = 6.0$ V. From Ohm's law,

$$I_7 = \frac{6.0\text{ V}}{8\ \Omega} = \frac{3}{4}\text{ A} \qquad I_{24} = \frac{6.0\text{ V}}{24\ \Omega} = \frac{1}{4}\text{ A}$$

As a check, 0.75 A + 0.25 A = 1.0 A which was the current I of the 6 Ω resistor. In Step 3, the 8 Ω resistor is returned to the 3 Ω and 5 Ω (right) resistors in series, so the two resistors must have the same current of 0.828 A. We use Ohm's law to find

$$\Delta V_3 = (3/4\ \text{A})(3\ \Omega) = 9/4\ \text{V} \qquad \Delta V_4 = (3/4\ \text{A})(5\ \Omega) = 15/4\ \text{V}$$

As a check, 9/4 V + 15/4 V = 24/4 V = 6.0 V, which was ΔV of the 8 Ω resistor. In Step 4, the 3 Ω resistor is returned to 4 Ω (left) and 12 Ω resistors in parallel, so the two must have the same potential difference ΔV = 9/4 V. From Ohm's law,

$$I_4 = \frac{9/4\ \text{V}}{4\ \Omega} = 9/16\ \text{A} = 0.56\ \text{A} \qquad I_{12} = \frac{9/4\ \text{V}}{12\ \Omega} = \frac{9}{48}\ \text{A} = 0.19\ \text{A}$$

As a check, 0.56 A + 0.19 A = 0.75 A, which was the same as the current through the 3 Ω resistor.

Resistor	Potential difference (V)	Current (A)
24 Ω	6.0	0.25
3 Ω	3.0	1.0
5 Ω	3.75	0.75
4 Ω	2.25	0.56
12 Ω	2.25	0.19

32.73. Model: The battery and the connecting wires are ideal.
Visualize: Please refer to Figure P32.73.
Solve: **(a)** A very long time after the switch has closed, the potential difference ΔV_C across the capacitor is $\mathcal{E}$. This is because the capacitor charges until $\Delta V_C = \mathcal{E}$, while the charging current approaches zero.
(b) The full charge of the capacitor is $Q_{max} = C(\Delta V_C)_{max} = C\mathcal{E}$.
(c) In this circuit, $I = +dQ/dt$ because the capacitor is charging, that is, because the charge on the capacitor is increasing.
(d) From Equation 32.36, capacitor charge at time t is $Q = Q_{max}(1 - e^{-t/\tau})$. So,

$$I = \frac{dQ}{dt} = C\mathcal{E}\frac{d}{dt}\left(1 - e^{-t/\tau}\right) = C\mathcal{E}\left(\frac{1}{\tau}\right)e^{-t/\tau} = C\mathcal{E}\left(\frac{1}{RC}\right)e^{-t/\tau} = \frac{\mathcal{E}}{R}e^{-t/\tau}$$

A graph of I as a function of t is shown below.

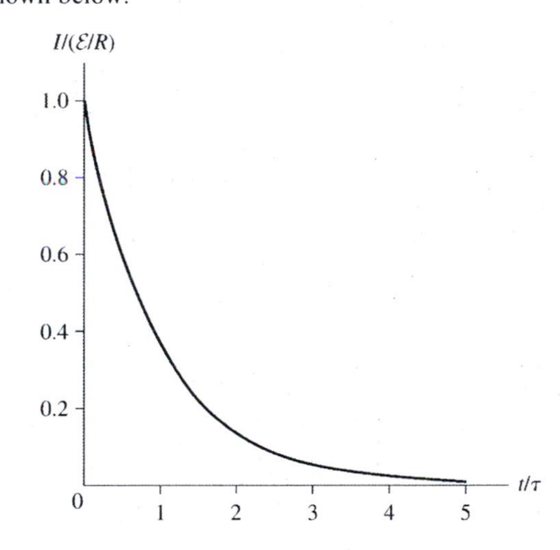

33

THE MAGNETIC FIELD

Exercises and Problems

33.3. Model: The magnetic field is that of a moving charged particle.
Visualize:

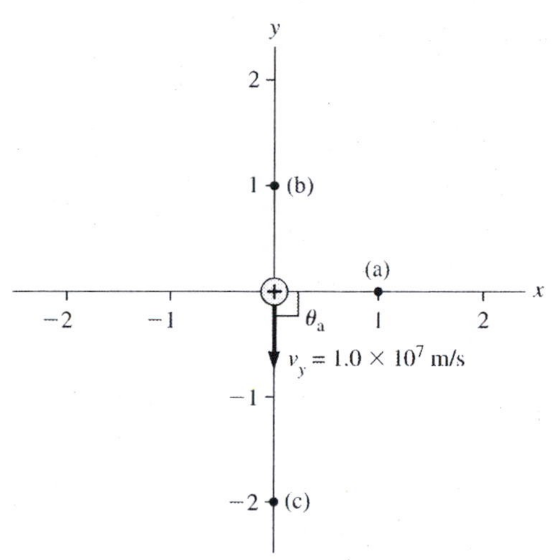

The first point is on the x-axis, with $\theta_a = 90°$. The second point is on the y-axis, with $\theta_b = 180°$, and the third point is on the $-y$-axis with $\theta_c = 0°$.
Solve: **(a)** Using Equation 33.1, the Biot-Savart law, the magnetic field strength is

$$B_a = \frac{\mu_0}{4\pi}\frac{qv\sin\theta}{r^2} = \frac{(10^{-7}\text{ T m/A})(1.60\times10^{-19}\text{ C})(1.0\times10^{+7}\text{ m/s})\sin 90°}{(1.0\times10^{-2}\text{ m})^2} = 1.60\times10^{-15}\text{ T}$$

To use the right-hand rule for finding the direction of $\vec{B}$, point your thumb in the direction of $\vec{v}$. The magnetic field vector $\vec{B}$ is perpendicular to the plane of $\vec{r}$ and $\vec{v}$ and points in the same direction that your fingers point. In the present case, the fingers point along the $\hat{k}$ direction. Thus, $\vec{B}_a = 1.60\times10^{-15}\hat{k}$ T.
(b) $B_b = 0$ T because $\sin\theta_b = \sin 180° = 0$.
(c) $B_c = 0$ T because $\sin\theta_c = \sin 0° = 0$.

33.9. Model: The magnetic field is that of an electric current in a long straight wire.
Solve: From Example 33.3, the magnetic field strength of a long, straight wire carrying current I at a distance d from the wire is

$$B = \frac{\mu_0}{2\pi}\frac{I}{d}$$

The distance d at which the magnetic field is equivalent to Earth's magnetic field is calculated as follows:

$$B_{\text{earth surface}} = 5\times10^{-5}\text{ T} = \left(2\times10^{-7}\text{ T m/A}\right)\frac{10\text{ A}}{d} \Rightarrow d = 4.0\text{ cm}$$

Likewise, the corresponding distances for a refrigerator magnet, a laboratory magnet, and a superconducting magnet are 0.40 mm, 20 μm to 2.0 μm, and 0.20 μm.

33.13. Model: Assume the wires are infinitely long.
Visualize:

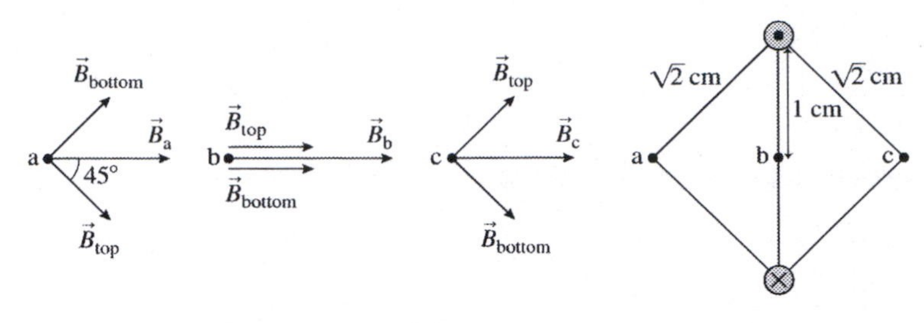

The field vectors are tangent to circles around the currents. The net magnetic field is the vectorial sum of the fields $\vec{B}_{\text{top}}$ and $\vec{B}_{\text{bottom}}$. Points a and c are at a distance $d = \sqrt{2}$ cm from both wires and point b is at a distance d = 1 cm.

Solve: The magnetic field at points a, b, and c are

$$\vec{B}_{\text{a}} = \vec{B}_{\text{top}} + \vec{B}_{\text{bottom}} = \frac{\mu_0 I}{2\pi d}\left(\cos 45°\hat{i} - \sin 45°\hat{j}\right) + \frac{\mu_0 I}{2\pi d}\left(\cos 45°\hat{i} + \sin 45°\hat{j}\right)$$

$$= \frac{\mu_0 I}{2\pi d} 2\cos 45°\hat{i} = \frac{\left(2\times10^{-7}\text{ T m/A}\right)(10\text{ A})}{\sqrt{2}\times10^{-2}\text{ m}} 2\left(\frac{1}{\sqrt{2}}\right)\hat{i} = 2.0\times10^{-4}\hat{i}\text{ T}$$

$$\vec{B}_{\text{b}} = \frac{\mu_0 I}{2\pi d}\hat{i} + \frac{\mu_0 I}{2\pi d}\hat{i} = 2\frac{\left(2\times10^{-7}\text{ T m/A}\right)(10\text{ A})}{1\times10^{-2}\text{ m}}\hat{i} = 4.0\times10^{-4}\hat{i}\text{ T}$$

$$\vec{B}_{\text{c}} = \frac{\mu_0 I}{2\pi d}\left(\cos 45°\hat{i} + \sin 45°\hat{j}\right) + \frac{\mu_0 I}{2\pi d}\left(\cos 45°\hat{i} - \sin 45°\hat{j}\right) = 2.0\times10^{-4}\hat{i}\text{ T}$$

33.15. Model: Assume that the 10 cm distance is much larger than the size of the small bar magnet.
Solve: **(a)** From Equation 33.9, the on-axis field of a magnetic dipole is

$$B = \frac{\mu_0}{4\pi}\frac{2\mu}{z^3} \Rightarrow \mu = \frac{4\pi}{\mu_0}\frac{Bz^3}{2} = \frac{\left(5.0\times10^{-6}\text{ T}\right)(0.10\text{ m})^3}{2\left(10^{-7}\text{ T m/A}\right)} = 0.025\text{ A m}^2$$

(b) The on-axis field strength 15 cm from the magnet is

$$B = \frac{\mu_0}{4\pi}\frac{2\mu}{z^3} = \left(10^{-7}\text{ T m/A}\right)\frac{2\left(0.025\text{ A m}^2\right)}{(0.15\text{ m})^3} = 1.48\times10^{-6}\text{ T} = 1.48\,\mu\text{T}$$

33.21. Model: The magnetic field is that of the three currents enclosed by the loop.
Visualize: Please refer to Figure EX33.21.
Solve: Ampere's law gives the line integral of the magnetic field around the closed path:

$$\oint \vec{B}\cdot d\vec{s} = \mu_0 I_{\text{through}} = 3.77\times10^{-6}\text{ T m} = \mu_0\left(I_1 - I_2 + I_3\right) = \left(4\pi\times10^{-7}\text{ T m/A}\right)\left(6.0\text{ A} - 4.0\text{ A} + I_3\right)$$

$$\Rightarrow \left(I_3 + 2.0\text{ A}\right) = \frac{3.77\times10^{-6}\text{ T m}}{4\pi\times10^{-7}\text{ T m/A}} \Rightarrow I_3 = 1.0\text{ A}$$

Assess: The right-hand rule was used above to assign positive signs to I_1 and I_3 and a negative sign to I_2.

33.27. Model: A magnetic field exerts a magnetic force on a moving charge.
Visualize: Please refer to Figure EX33.27.
Solve: **(a)** The force is

$$\vec{F}_{\text{on q}} = q\vec{v}\times\vec{B} = \left(-1.60\times10^{-19}\text{ C}\right)\left(-1.0\times10^{7}\hat{j}\text{ m/s}\right)\times\left(0.50\hat{i}\text{ T}\right) = -8.0\times10^{-13}\hat{k}\text{ N}$$

(b) The force is

$$\vec{F}_{\text{on q}} = \left(-1.60\times10^{-19}\text{ C}\right)\left(1.0\times10^{7}\text{ m/s}\right)\left(-\cos45°\hat{j}+\sin45°\hat{k}\right)\times\left(0.50\hat{i}\text{ T}\right) = 5.7\times10^{-13}\left(-\hat{j}-\hat{k}\right)\text{ N}$$

33.33. Model: Assume the magnetic field is uniform over the Hall probe.
Visualize: Please refer to Figure EX33.42(a). The thickness is $t = 4.0 \times 10^{-3}$ m.
Solve: The Hall voltage is given by Equation 33.24:

$$\Delta V_{\text{H}} = \frac{IB}{tne} \Rightarrow n = \frac{IB}{te\Delta V_{\text{H}}} = \frac{(15\text{ A})(1.0\text{ T})}{\left(1.0\times10^{-3}\text{ m}\right)\left(1.60\times10^{-19}\text{ C}\right)\left(3.2\times10^{-6}\text{ V}\right)} = 2.9\times10^{28}\text{ m}^{-3}$$

Assess: The conduction electron density in metals is of the order of $\approx 5 \times 10^{+28}$ m^{-3} (Table 31.1). The value obtained for the charge carrier density is reasonable.

33.37. Model: Two parallel wires carrying currents in the same direction exert attractive magnetic forces on each other.
Visualize: Please refer to Figure EX33.37. The current in the circuit on the left is I_1 and has a clockwise direction. The current in the circuit on the right is I_2 and has a counterclockwise direction.
Solve: Since $I_1 = 9\text{ V}/2\ \Omega = 4.5$ A, the force between the two wires is

$$F = 5.4\times10^{-5}\text{ N} = \frac{\mu_0 L I_1 I_2}{2\pi d} = \frac{\left(2\times10^{-7}\text{ T m/A}\right)(0.10\text{ m})(4.5\text{ A})I_2}{0.0050\text{ m}}$$

$$\Rightarrow I_2 = 3.0\text{ A} \Rightarrow R = \frac{9\text{ V}}{3.0\text{ A}} = 3.0\ \Omega$$

33.41. Model: The torque on the current loop is due to the magnetic field produced by the current-carrying wire. Assume that the wire is very long.
Visualize: Please refer to Figure EX33.41.
Solve: **(a)** From Equation 33.27, the magnitude of the torque on the current loop is $\tau = \mu B\sin\theta$, where $\mu = I_{\text{loop}}A$ and B is the magnetic field produced by the current I_{wire} in the wire. The magnetic field of the wire is tangent to a circle around the wire. At the position of the loop, $\vec{B}$ points up and is $\theta = 90°$ from the axis of the loop. Thus,

$$\tau = \left(I_{\text{loop}}A\right)\frac{\mu_0 I_{\text{wire}}}{2\pi d}\sin\theta = \frac{(0.20\text{ A})\pi(0.0010\text{ m})^2\left(2\times10^{-7}\text{ T m/A}\right)(2.0\text{ A})\sin90°}{2.0\times10^{-2}\text{ m}} = 1.26\times10^{-11}\text{ N m}$$

Note that the magnetic field produced by the wire on the current loop is *up* so that the angle θ between $\vec{B}$ and the normal to the loop is 90°.
(b) The loop is in equilibrium when $\theta = 0°$ or 180°. That is, when the coil is rotated by $\pm 90°$.

33.43. Model: Assume that the wires are infinitely long and that the magnetic field is due to currents in both the wires.
Visualize: Point 1 is a distance d_1 away from the two wires and point 2 is a distant d_2 away from the two wires. A right triangle with a 75° degree angle is formed by a straight line from point 1 to the intersection and a line from point 1 that is perpendicular to the wire. Likewise, point 2 makes a 15° right triangle.

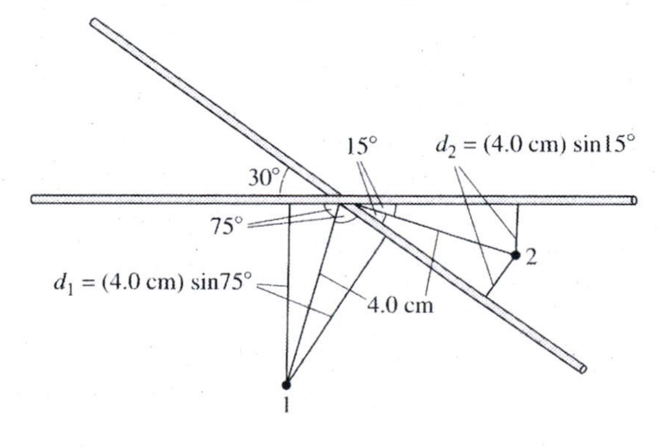

Solve: First we determine the distances d_1 and d_2 of the points from the two wires:

$$d_1 = (4.0 \text{ cm})\sin 75° = 3.86 \text{ cm} = 0.0386 \text{ m}$$

$$d_2 = (4.0 \text{ cm})\sin 15° = 1.04 \text{ cm} = 0.0104 \text{ m}$$

At point 1, the fields from both the wires point up and hence add. The total field is

$$B_1 = B_{\text{wire 1}} + B_{\text{wire 2}} = \frac{\mu_0}{2\pi}\frac{I_1}{d_1} + \frac{\mu_0}{2\pi}\frac{I_2}{d_1} = \frac{\mu_0}{\pi}\frac{(5.0 \text{ A})}{d_1} = \frac{(4\times10^{-7} \text{ T m/A})(5.0 \text{ A})}{0.0386 \text{ m}} = 5.2\times10^{-5} \text{ T}$$

In vector form, $\vec{B}_1$ = (5.2 × 10^{-5} T, out of page). Using the right-hand rule at point 2, the fields are in opposite directions but equal in magnitude. So, $\vec{B}_2 = \vec{0}$ T.

33.47. Model: Use the Biot-Savart law for a current carrying segment.
Visualize: Please refer to Figure P33.47. The distance from P to the inner arc is r_1 and the distance from P to the outer arc is r_2.
Solve: As given in Equation 33.6, the Biot-Savart law for a current carrying small segment $\Delta\vec{s}$ is

$$\vec{B} = \frac{\mu_0}{4\pi}\frac{I\Delta\vec{s}\times\hat{r}}{r^2}$$

For the linear segments of the loop, $B_{\Delta s}$ = 0 T because $\Delta\vec{s}\times\hat{r} = 0$. Consider a segment $\Delta\vec{s}$ on length on the inner arc. Because $\Delta\vec{s}$ is perpendicular to the $\hat{r}$ vector, we have

$$B = \frac{\mu_0}{4\pi}\frac{I\Delta s}{r_1^2} = \frac{\mu_0}{4\pi}\frac{Ir_1\Delta\theta}{r_1^2} = \frac{\mu_0}{4\pi}\frac{I\Delta\theta}{r_1} \Rightarrow B_{\text{arc 1}} = \int_{-\pi/2}^{\pi/2}\frac{\mu_0 I d\theta}{4\pi r_1} = \frac{\mu_0 I}{4\pi r_1}\pi = \frac{\mu_0 I}{4r_1}$$

A similar expression applies for $B_{\text{arc 2}}$. The right-hand rule indicates an out-of-page direction for $B_{\text{arc 2}}$ and an into-page direction for $B_{\text{arc 1}}$. Thus,

$$\vec{B} = \left(\frac{\mu_0 I}{4r_1}, \text{ into page}\right) + \left(\frac{\mu_0 I}{4r_2}, \text{ out of page}\right) = \left[\frac{\mu_0 I}{4}\left(\frac{1}{r_1} - \frac{1}{r_2}\right), \text{ into page}\right]$$

The field strength is

$$B = \frac{(4\pi\times10^{-7} \text{ T m/A})(5.0 \text{ A})}{4}\left(\frac{1}{0.010 \text{ m}} - \frac{1}{0.020 \text{ m}}\right) = 7.9\times10^{-5} \text{ T}$$

Thus $\vec{B}$ = (7.9 × 10^{-5} T, into page).

33.51. Model: A 1000-km-diameter ring makes a loop of diameter 3000 km.
Visualize:

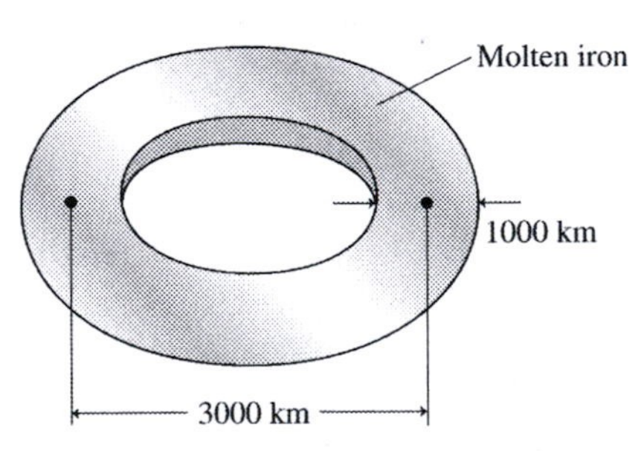

Solve: **(a)** The current loop has a diameter of 3000 km, so its nominal area, ignoring curvature effects, is

$$A_{\text{loop}} = \pi r^2 = \pi(1500 \times 10^3 \text{ m})^2 = 7.07 \times 10^{12} \text{ m}^2$$

Because the magnetic dipole moment of the earth is modeled to be due to a current flowing in such a loop, $\mu = IA_{\text{loop}}$. The current in the loop is

$$I = \frac{\mu}{A_{\text{loop}}} = \frac{8.0\times10^{22} \text{ A m}^2}{7.07\times10^{12} \text{ m}^2} = 1.13\times10^{10} \text{ A}$$

(b) The current density J in the above loop is

$$J_{\text{loop}} = \frac{I}{A} = \frac{1.13 \times 10^{10}\ \text{A}}{\pi\left(\frac{1}{2} \times 1000 \times 10^{3}\ \text{m}\right)^{2}} = 0.014\ \text{A/m}^{2}$$

(c) The current density in the wire is

$$J_{\text{wire}} = \frac{I}{A} = \frac{1.0\ \text{A}}{\pi\left(\frac{1}{2} \times 1.0 \times 10^{-3}\ \text{m}\right)^{2}} = 1.3 \times 10^{6}\ \text{A/m}^{2}$$

You can see that $J_{\text{loop}} << J_{\text{wire}}$. The current in the earth's core is large, but the current density is actually quite small.

33.53. Model: The magnetic field is that of a current in the wire.
Visualize: Please refer to Figure P33.53.
Solve: As given in Equation 33.6 for a current carrying small segment $\Delta\vec{s}$, the Biot-Savart law is

$$\vec{B} = \frac{\mu_0}{4\pi}\frac{I\Delta\vec{s} \times \hat{r}}{r^2}$$

For the straight sections, $\Delta\vec{s} \times \hat{r} = 0$ because both $\Delta\vec{s}$ and $\hat{r}$ point along the same line. That is not the case with the curved section over which $\Delta\vec{s}$ and $\vec{r}$ are perpendicular. Thus,

$$B = \frac{\mu_0}{4\pi}\frac{I\Delta s}{r^2} = \frac{\mu_0}{4\pi}\frac{IR\,d\theta}{R^2} = \frac{\mu_0 I\,d\theta}{4\pi R}$$

where we used $\Delta s = R\Delta\theta \approx R\,d\theta$ for the small arc length Δs. Integrating to obtain the total magnetic field at the center of the semicircle,

$$B = \int_{-\pi/2}^{\pi/2}\frac{\mu_0 I\,d\theta}{4\pi R} = \frac{\mu_0 I}{4\pi R}\pi = \frac{\mu_0 I}{4R}$$

33.55. Model: The magnetic field is that of the current which is distributed uniformly in the hollow wire.
Visualize:

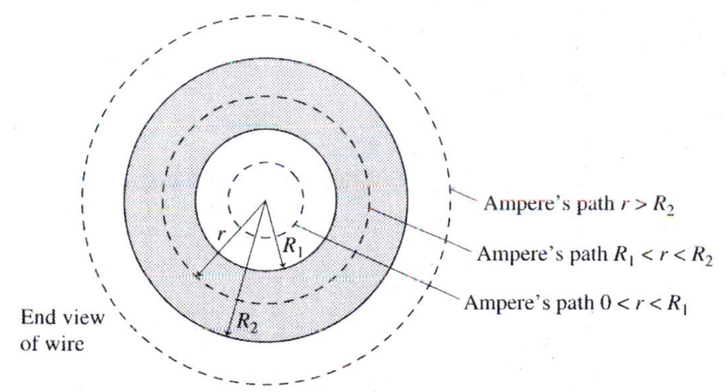

Ampere's integration paths are shown in the figure for the regions $0\ \text{m} < r < R_1$, $R_1 < r < R_2$, and $R_2 < r$.
Solve: For the region $0\ \text{m} < r < R_1$, $\oint\vec{B}\cdot d\vec{s} = \mu_0 I_{\text{through}}$. Because the current inside the integration path is zero, $B = 0$ T. To find I_{through} in the region $R_1 < r < R_2$, we multiply the current density by the area inside the integration path that carries the current. Thus,

$$I_{\text{through}} = \frac{I}{\pi\left(R_2^2 - R_1^2\right)}\pi\left(r^2 - R_1^2\right)$$

where the current density is the first term. Because the magnetic field has the same magnitude at every point on the circular path of integration, Ampere's law simplifies to

$$\oint\vec{B}\cdot d\vec{s} = B\int ds = B(2\pi r) = \mu_0\frac{I\left(r^2 - R_1^2\right)}{\left(R_2^2 - R_1^2\right)} \Rightarrow B = \frac{\mu_0 I}{2\pi r}\left(\frac{r^2 - R_1^2}{R_2^2 - R_1^2}\right)$$

For the region $R_2 < r$, $I_{through}$ is simply I because the loop encompasses the entire current. Thus,

$$\oint \vec{B} \cdot d\vec{s} = B \int ds = B 2\pi r = \mu_0 I \Rightarrow B = \frac{\mu_0 I}{2\pi r}$$

Assess: The results obtained for the regions $r > R_2$ and $R_1 < r < R_2$ yield the same result at $r = R_2$. Also note that a hollow wire and a regular wire have the same magnetic field outside the wire.

33.59. Model: Energy is conserved as the electron moves between the two electrodes. Assume the electron starts from rest. Once in the magnetic field, the electron moves along a circular arc.
Visualize:

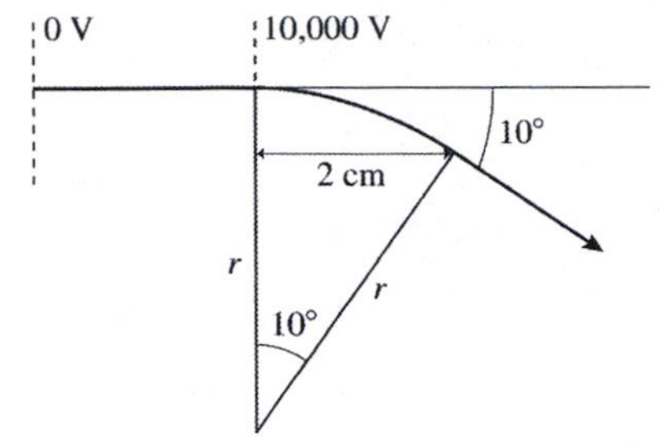

The electron is deflected by 10° after moving along a circular arc of angular width 10°.
Solve: Energy is conserved as the electron moves from the 0 V electrode to the 10,000 V electrode. The potential energy is $U = qV$ with $q = -e$, so

$$K_f + U_f = K_i + U_i \Rightarrow \tfrac{1}{2}mv^2 - eV = 0 + 0$$

$$v = \sqrt{\frac{2eV}{m}} = \sqrt{\frac{2(1.60\times10^{-19}\text{ C})(10{,}000\text{ V})}{9.11\times10^{-31}\text{ kg}}} = 5.93\times10^{7}\text{ m/s}$$

The radius of cyclotron motion in a magnetic field is $r = mv/eB$. From the figure we see that the radius of the circular arc is $r = (2.0\text{ cm})/\sin10°$. Thus

$$B = \frac{mv}{er} = \frac{(9.11\times10^{-31}\text{ kg})(5.93\times10^{7}\text{ m/s})}{(1.60\times10^{-19}\text{ C})(0.020\text{ m})/\sin10°} = 2.9\times10^{-3}\text{ T}$$

33.63. Model: Charged particles moving perpendicular to a uniform magnetic field undergo uniform circular motion at constant speed.
Solve: **(a)** The magnetic force on a proton causes a centripetal acceleration:

$$evB = \frac{mv^2}{r} \Rightarrow v = \frac{eBr}{m}$$

Maximum kinetic energy is achieved when the diameter of the proton's orbit matches the diameter of the cyclotron:

$$K = \tfrac{1}{2}mv^2 = \frac{e^2B^2r^2}{2m} = \frac{\left(1.60\times10^{-19}\text{ C}\right)^2\left(0.75\text{ T}\right)^2\left(0.325\text{ m}\right)^2}{2\left(1.67\times10^{-27}\text{ kg}\right)} = 4.6\times10^{-13}\text{ J}$$

(b) The proton accelerates through a potential difference of 500 V twice during one revolution. The energy gained per cycle is

$$2\,q\Delta V = 2e\,(500\text{ V}) = 1.60\times10^{-16}\text{ J}$$

Using the maximum kinetic energy of the proton from part (a), the number of cycles before the proton attains this energy is

$$\frac{4.6\times10^{-13}\text{ J}}{1.60\times10^{-16}\text{ J}} = 2850$$

33.65. Model: Assume that the magnetic field is uniform over the Hall probe.
Solve: Equation 33.24 gives the Hall voltage and Equation 33.20 gives the cyclotron frequency in terms of the magnetic field. We have

$$\Delta V_{\text{H}} = \frac{IB}{tne} \qquad B = \frac{2\pi m f_{\text{cyl}}}{e}$$

$$\Rightarrow tne = \frac{2\pi m f_{\text{cyc}} I}{e\Delta V_{\text{H}}} = \frac{2\pi\left(1.67\times10^{-27}\text{ kg}\right)\left(10.0\times10^{6}\text{ Hz}\right)\left(0.150\times10^{-3}\text{ A}\right)}{\left(1.60\times10^{-19}\text{ C}\right)\left(0.543\times10^{-3}\text{ V}\right)} = 0.1812\text{ T A/V}$$

With this value of *tne*, we can once again use the Hall voltage equation to find the magnetic field:

$$B = \left(\frac{\Delta V_{\text{H}}}{I}\right)tne = \left(\frac{1.735\times10^{-3}\text{ V}}{0.150\times10^{-3}\text{ A}}\right)(0.1812\text{ TA/V}) = 2.10\text{ T}$$

33.69. Model: A magnetic field exerts a magnetic force on a length of current-carrying wire.
Visualize:

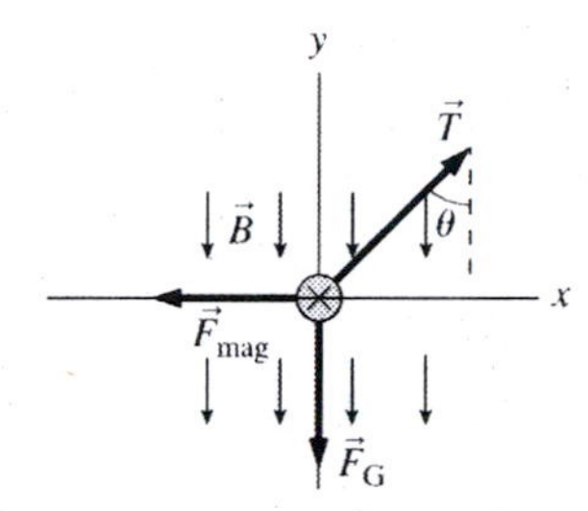

Please refer to Figure P33.69.
Solve: The above figure shows a side view of the wire, with the current moving into the page. From the right-hand rule, the magnetic field $\vec{B}$ points *down* to give a leftward force on the current. The wire is hanging in static equilibrium, so $\vec{F}_{\text{net}} = \vec{F}_{\text{mag}} + \vec{F}_G + \vec{T} = 0$ N. Consider a segment of wire of length L. The wire's linear mass density is $\mu = 0.050$ kg/m, so the mass of this segment is $m = \mu L$ and its weight is $F_G = mg = \mu Lg$. The magnetic force on this length of wire is $F_{\text{mag}} = ILB$. In component form, Newton's first law is

$$(F_{\text{net}})_x = T\sin\theta - F_{\text{mag}} = T\sin\theta - ILB = 0\text{ N} \Rightarrow T\sin\theta = ILB$$

$$(F_{\text{net}})_y = T\cos\theta - F_G = T\cos\theta - \mu Lg = 0\text{ N} \Rightarrow T\cos\theta = \mu Lg$$

Dividing the first equation by the second,

$$\left[\frac{T\sin\theta}{T\cos\theta} = \tan\theta\right] = \left[\frac{ILB}{\mu Lg} = \frac{IB}{\mu g}\right] \Rightarrow B = \frac{\mu g\tan\theta}{I} = \frac{(0.050\text{ kg/m})(9.8\text{ m/s}^2)\tan 10°}{10\text{ A}} = 0.0086\text{ T}$$

The magnetic field is $\vec{B} = (8.6\text{ mT, down})$.

34

ELECTROMAGNETIC INDUCTION

Exercises and Problems

34.3. Visualize:

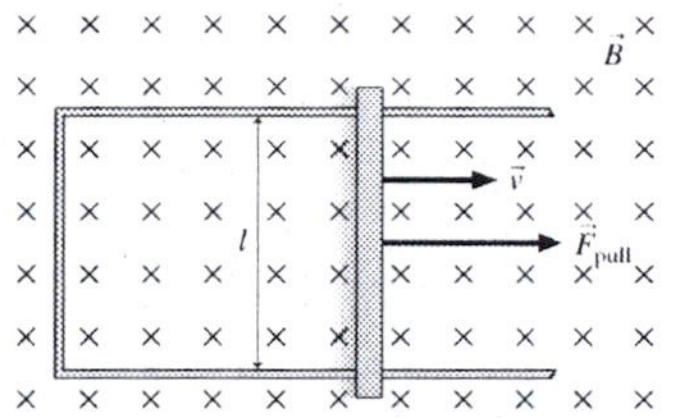

The wire is pulled with a constant force in a magnetic field. This results in a motional emf and produces a current in the circuit. From energy conservation, the mechanical power provided by the puller must appear as electrical power in the circuit.
Solve: (a) Using Equation 34.6,

$$P = F_{\text{pull}} v \Rightarrow v = \frac{P}{F_{\text{pull}}} = \frac{4.0 \text{ W}}{1.0 \text{ N}} = 4.0 \text{ m/s}$$

(b) Using Equation 34.6 again,

$$P = \frac{v^2 l^2 B^2}{R} \Rightarrow B = \sqrt{\frac{RF_{\text{pull}}}{vl^2}} = \sqrt{\frac{(0.20\ \Omega)(1.0 \text{ N})}{(4.0 \text{ m/s})(0.10 \text{ m})^2}} = 2.2 \text{ T}$$

Assess: This is reasonable field for the circumstances given.

34.5. Model: Consider the solenoid to be long so the field is constant inside and zero outside.
Visualize: Please refer to Figure Ex34.5. The field of a solenoid is along the axis. The flux through the loop is only nonzero inside the solenoid. Since the loop completely surrounds the solenoid, the total flux through the loop will be the same in both the perpendicular and tilted cases.

Solve: The field is constant inside the solenoid so we will use Equation 34.10. Take $\vec{A}$ to be in the same direction as the field. The magnetic flux is

$$\Phi = \vec{A}_{\text{loop}} \cdot \vec{B}_{\text{loop}} = \vec{A}_{\text{sol}} \cdot \vec{B}_{\text{sol}} = \pi r_{\text{sol}}^2 B_{\text{sol}} \cos\theta = \pi (0.010 \text{ m})^2 (0.20 \text{ T}) = 6.3 \times 10^{-5} \text{ Wb}$$

When the loop is tilted the component of $\vec{B}$ in the direction of $\vec{A}$ is less, but the effective area of the loop surface through which the magnetic field lines cross is increased by the same factor.

34.9. Visualize: Please refer to Figure Ex34.9. The changing current in the solenoid produces a changing flux in the loop. By Lenz's law there will be an induced current and field to oppose the change in flux.
Solve: The current shown produces a field to the right inside the solenoid. So there is flux to the right through the surrounding loop. As the current in the solenoid increases there is more field and more flux to the right through the loop. There is an induced current in the loop that will oppose the *change* by creating an induced field and flux to the left. This requires a *counterclockwise* current.

34.13. Model: Assume the field strength is changing at a constant rate.
Visualize:

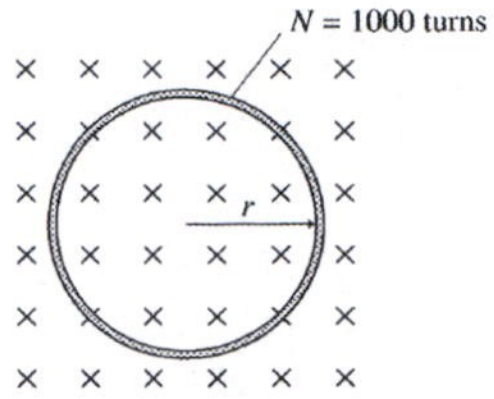

The changing field produces a changing flux in the coil and there will be a corresponding induced emf and current.
Solve: The induced emf of the coil is

$$\mathcal{E} = N\left|\frac{d\Phi}{dt}\right| = N\left|\frac{d\left(\vec{A}\cdot\vec{B}\right)}{dt}\right| = NA\left|\frac{dB}{dt}\right| = N\pi r^2\left|\frac{dB}{dt}\right| = \left(10^3\right)\pi\left(0.010\text{ m}\right)^2\left(\frac{0.10\text{ T}}{10\times10^{-3}\text{ s}}\right) = 3.1\text{ V}$$

where we've used the fact that $\vec{B}$ is parallel to $\vec{A}$.
Assess: This seems to be a reasonable emf as there are many turns.

34.19. Model: Assume that the current changes uniformly.
Visualize: We want to increase the current without exceeding a maximum potential difference.
Solve: Since we want the minimum time, we will use the maximum potential difference:

$$\Delta V = -L\frac{dI}{dt} = -L\frac{\Delta I}{\Delta t} \Rightarrow \Delta t = L\left|\frac{\Delta I}{\Delta V_{\text{max}}}\right| = \left(200\times10^{-3}\text{ H}\right)\frac{3.0\text{ A} - 1.0\text{ A}}{400\text{ V}} = 1.0\text{ ms}$$

Assess: If we change the current in any shorter time the potential difference will exceed the limit.

34.23. Visualize: Changing the variable capacitor in combination with a fixed inductor will change the resonant frequency of the *LC* circuit.
Solve: Since the resonant frequency depends on the inverse square root of the capacitance a lower capacitance will produce a higher frequency and vice versa. The maximum frequency is

$$\omega_{\text{max}} = \sqrt{\frac{1}{LC_{\text{min}}}} = \sqrt{\frac{1}{\left(2.0\times10^{-3}\text{ H}\right)\left(100\times10^{-12}\text{ F}\right)}} = 2.2\times10^6\text{ rad/s}$$

The corresponding minimum is $\omega_{\text{min}} = 2.0\times10^6$ rad/s. These are angular frequencies so we can use $f = \omega/2\pi$ to find $f_{\text{min}} = 2.5\times10^5$ Hz and $f_{\text{max}} = 3.6\times10^5$ Hz, giving a range of 250 kHz to 360 kHz.

34.25. Visualize: Please refer to Figure Ex34.25. This is a simple *LR* circuit if the resistors in parallel are treated as an equivalent resistor in series with the inductor.
Solve: We can find the equivalent resistance from the time constant since we know the inductance. We have

$$\tau = \frac{L}{R_{\text{eq}}} \Rightarrow R_{\text{eq}} = \frac{L}{\tau} = \frac{3.6\times10^{-3}\text{ H}}{10\times10^{-6}\text{ s}} = 360\ \Omega$$

The equivalent resistance is the parallel addition of the unknown resistor *R* and 600 Ω. We have

$$\frac{1}{R_{\text{eq}}} = \frac{1}{600\ \Omega} + \frac{1}{R} \Rightarrow R = \frac{(600\ \Omega)(360\ \Omega)}{600\ \Omega - 360\ \Omega} = 900\ \Omega$$

34.29. Model: Assume the field is uniform in space though it is changing in time.
Visualize: The changing magnetic field strength produces a changing flux through the loop, and a corresponding induced emf and current.
Solve: **(a)** Since the field is perpendicular to the plane of the loop, $\vec{A}$ is parallel to $\vec{B}$ and $\Phi = AB$. The emf is

$$\mathcal{E} = \left|\frac{d\Phi}{dt}\right| = A\left|\frac{dB}{dt}\right| = (0.20\text{ m})^2\left|(4-4t)\text{T/s}\right| = \left|0.16(1-t)\right|\text{ V} \Rightarrow I = \frac{\mathcal{E}}{R} = \left|1.6(1-t)\right|\text{ A}$$

The magnetic field is increasing over the interval $0 \text{ s} < t < 1 \text{ s}$ and is decreasing over the interval $1 \text{ s} < t < 2 \text{ s}$, so the induced emf and current must have opposite signs in the second half of the time interval. We arbitrarily choose the sign to be positive during the first half.

Time (s)	B (T)	$\mathcal{E}$ (volts)	I (A)
0.0	0.00	0.16	1.6
0.5	1.50	0.08	0.8
1.0	2.00	0.00	0.0
1.5	1.50	–0.08	–0.8
2.0	0.00	–0.16	–1.6

(b) To plot the field and current we look at the form of the equations as a function of time. The magnetic field strength is quadratic with a maximum at $t = 1$ s and vanishing at $t = 0$ s and $t = 2$ s. The current equation is linear and decreasing, starting at 1.6 A at $t = 0$ s and going through zero at $t = 1$ s.

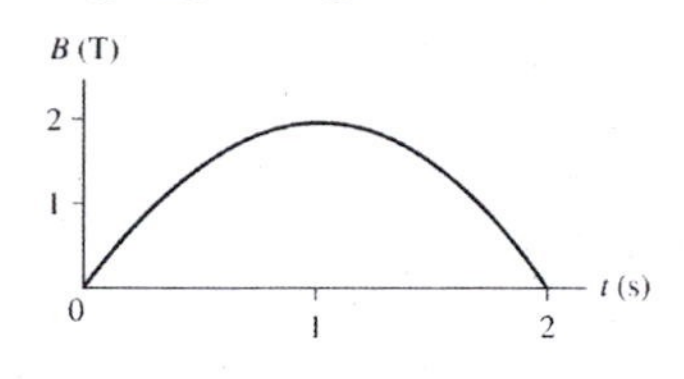

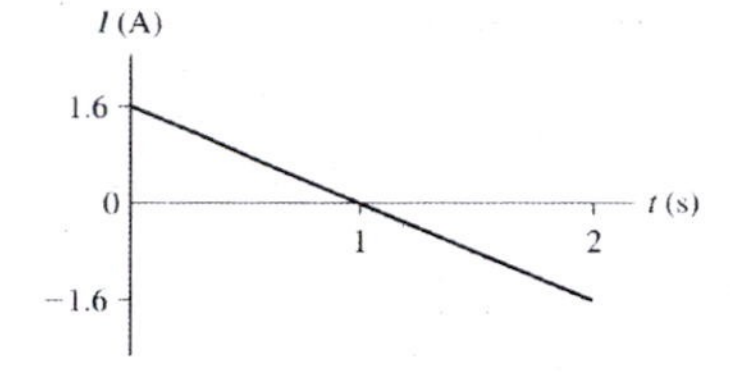

Assess: Notice in the graph how $I = 0$ A at $t = 1$ s, the instant in time when B is a maximum, that is, when $dB/dt = 0$. At this point the flux is (instantaneously) not changing so the corresponding induced emf and current are zero.

34.37. Model: Assume the wire is long enough so we can use the formula for the magnetic field of an "infinite" wire.
Visualize:

The magnetic field in the vicinity of the loop is due to the current in the wire and is perpendicular to the loop. The current is changing so the field and the flux through the loop are changing. This will create an induced emf and induced current in the loop.
Solve: The induced current depends on the induced emf and is

$$I_{\text{loop}} = \frac{\mathcal{E}_{\text{loop}}}{R} = \frac{1}{R}\left|\frac{d\Phi}{dt}\right|$$

The flux through a rectangular loop due to a wire was found in Example 34.5. The total flux is

$$\Phi = \frac{\mu_0 I b}{2\pi}\ln\left(\frac{c+a}{c}\right)$$

$$\Rightarrow I_{\text{loop}} = \frac{1}{R}\frac{\mu_0 b}{2\pi}\ln\left(\frac{c+a}{c}\right)\frac{dI}{dt} = \frac{(4\pi\times10^{-7}\text{ T m/A})\,(0.020\text{ m})}{(0.010\ \Omega)2\pi}\ln\left(\frac{0.030\text{ m}}{0.010\text{ m}}\right)(100\text{ A/s}) = 44\ \mu\text{A}$$

34.41. Model: Assume that the magnetic field of coil 1 passes through coil 2 and that we can use the magnetic field of a solenoid for coil 1.
Visualize: Please refer to Figure P34.41. The field of coil 1 produces flux in coil 2. The changing current in coil 1 gives a changing flux in coil 2 and a corresponding induced emf and current in coil 2.
Solve: **(a)** From 0 s to 0.1 s and 0.3 s to 0.4 s the current in coil 1 is constant so the current in coil 2 is zero. Thus $I(0.05\text{ s}) = 0$ A.

(b) From 0.1 s to 0.3 s, the induced current from the induced emf is given by Faraday's law. The current in coil 2 is

$$I_2 = \frac{\mathcal{E}_2}{R} = \frac{1}{R}N_2\left|\frac{d\Phi_2}{dt}\right| = \frac{1}{R}N_2A_2\left|\frac{dB_1}{dt}\right| = \frac{1}{R}N_2\pi r_2^2\left|\frac{d}{dt}\left(\frac{\mu_0 N_1 I_1}{l_1}\right)\right| = \frac{N_2\pi r_2^2\mu_0 N_1}{Rl_1}\left|\frac{dI_1}{dt}\right|$$

$$= \frac{20\pi(0.010\text{ m})^2(4\pi\times10^{-7}\text{ T m/A})(20)}{(2\Omega)(0.020\text{ m})}|20\text{ A/s}| = 7.95\times10^{-5}\text{ A} = 79\mu\text{A}$$

We used the facts that the field of coil 1 is constant inside the loops of coil 2 and the flux is confined to the area $A_2 = \pi r_2^2$ of coil 2. Also, we used $l_1 = N_1 d = 20(1.0\text{ mm}) = 0.020\text{ m}$ and $|dI/dt| = 20$ A/s. From 0.1 s to 0.2 s the current in coil 1 is initially negative so the field is initially to the right and the flux is decreasing. The induced current will *oppose this change* and will therefore produce a field to the right. This requires an induced current in coil 2 that comes out of the page at the top of the loops so it is negative. From 0.2 s to 0.3 s the current in coil 1 is positive so the field is to the left and the flux is increasing. The induced current will *oppose this change* and will therefore produce a field to the right. Again, this is a negative current. Hence $I(0.25\text{ s}) = 79\ \mu$A right to left through the resistor.

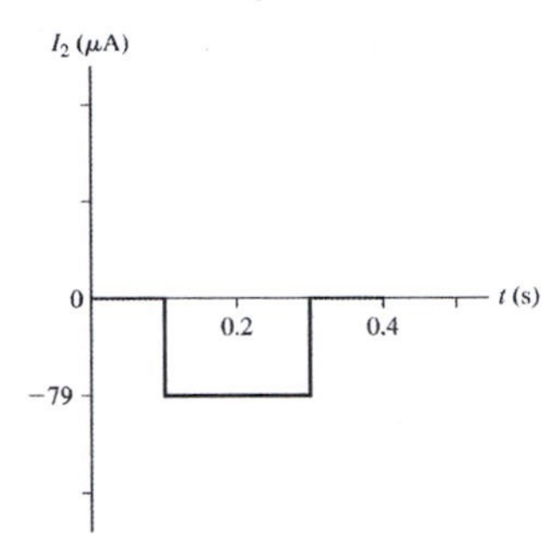

34.45. Model: Assume an ideal transformer.
Visualize: An ideal transformer changes the voltage, but not the power (energy conservation).
Solve: **(a)** The primary and secondary voltages are related by Equation 34.30. We have

$$V_2 = \frac{N_2}{N_1}V_1 \Rightarrow N_1 = \frac{V_1}{V_2}N_2 = \frac{15{,}000}{120}100 = 12{,}500\text{ turns}$$

(b) The input power equals the output power and we recall that $P = I\Delta V$, so

$$P_{\text{out}} = P_{\text{in}} \Rightarrow I_1\Delta V_1 = I_2\Delta V_2 \Rightarrow I_1 = \frac{I_2\Delta V_2}{\Delta V_1} = \frac{(250\text{ A})120\text{ V}}{15{,}000\text{ V}} = 2.0\text{ A}$$

Assess: These values seem reasonable, because houses have low voltage and high current while transmission lines have high voltage and low current.

34.51. Model: Assume that the magnetic field is uniform in the region of the loop.
Visualize: Please refer to Figure P34.51. The rotating semicircle will change the area of the loop and therefore the flux through the loop. This changing flux will produce an induced emf and corresponding current in the bulb.
Solve: **(a)** The spinning semicircle has a normal to the surface that changes in time, so while the magnetic field is constant, the area is changing. The flux through in the lower portion of the circuit does not change and will not contribute to the emf. Only the flux in the part of the loop containing the rotating semicircle will change. The flux associated with the semicircle is

$$\Phi = \vec{A}\cdot\vec{B} = BA = BA\cos\theta = BA\cos(2\pi ft)$$

where $\theta = 2\pi ft$ is the angle between the normal of the rotating semicircle and the magnetic field and A is the area of the semicircle. The induced current from the induced emf is given by Faraday's law. We have

$$I = \frac{\mathcal{E}}{R} = \frac{1}{R}\left|\frac{d\Phi}{dt}\right| = \frac{1}{R}\left|\frac{d}{dt}BA\cos(2\pi ft)\right| = \frac{B}{R}\frac{\pi r^2}{2}2\pi f\sin(2\pi ft)$$

$$= \frac{2(0.20\text{ T})\pi^2(0.050\text{ m})^2}{2(1.0\Omega)}f\sin(2\pi ft) = 4.9\times10^{-3}f\sin(2\pi ft)\text{ A}$$

where the frequency f is in Hz.

(b) We can now solve for the frequency necessary to achieve a certain current. From our study of DC circuits we know how power relates to resistance:

$$P = I^2 R \Rightarrow I = \sqrt{P/R} = \sqrt{4.0\text{ W}/1.0\Omega} = 2.0\text{ A}$$

The maximum of the sine function is +1, so the maximum current is

$$I_{\text{max}} = 4.9\times10^{-3} f \text{ A s} = 2.0\text{ A} \Rightarrow f = \frac{2.0\text{ A}}{4.9\times10^{-3}\text{ A s}} = 4.1\times10^2\text{ Hz}$$

Assess: This is not a reasonable frequency to obtain by hand.

34.53. Model: Assume the magnetic field is uniform in the region of the loop.
Visualize:

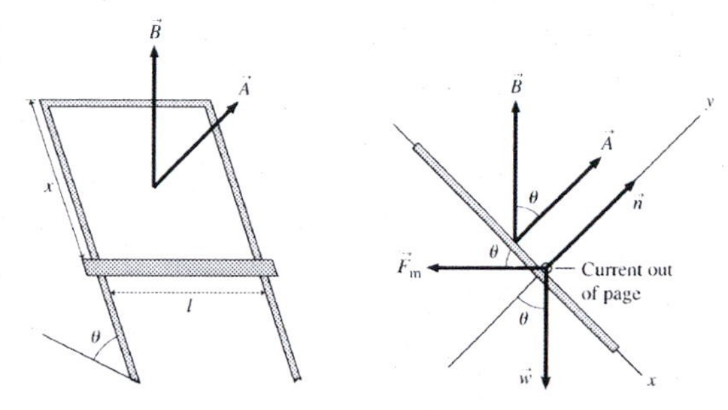

The moving wire creates a changing area and corresponding change in flux. This produces an induced emf and induced current. The flux through the loop depends on the size and orientation of the loop.

Solve: **(a)** The normal to the surface is perpendicular to the loop and the flux is $\Phi_{\text{inner}} = \vec{A}\cdot\vec{B} = AB\cos\theta$. We can get the current from Faraday's law. Since the loop area is $A = lx$, We have

$$I = \frac{\mathcal{E}}{R} = \frac{1}{R}\left|\frac{d\Phi}{dt}\right| = \frac{1}{R}\left|\frac{d}{dt}lxB\cos\theta\right| = \frac{Bl\cos\theta}{R}\left|\frac{dx}{dt}\right| = \frac{Blv\cos\theta}{R}$$

(b) Using the free-body diagram shown in the figure, we can apply Newton's second law. The magnetic force on a straight, current-carrying wire is $F_{\text{m}} = IlB$ and is horizontal. Using the current I from part (a) gives

$$\sum F_x = -F_{\text{m}}\cos\theta + mg\sin\theta = -\frac{B^2l^2v\cos^2\theta}{R} + mg\sin\theta = ma_x$$

Terminal speed is reached when a_x drops to zero. In this case, the two terms are equal and we have

$$v_{\text{term}} = \frac{mgR\tan\theta}{l^2B^2\cos\theta}$$

34.57. Model: Assume the field is uniform in the region of the coil.
Visualize:

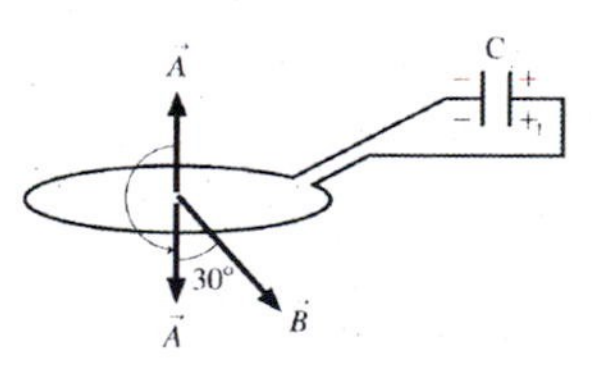

The rotation of the coil in the field will change the flux and produce an induced emf and a corresponding induced current. The current will charge the capacitor.

Solve: The induced current is

$$I_{\text{coil}} = \frac{\mathcal{E}_{\text{coil}}}{R} = \frac{N}{R}\left|\frac{d\Phi}{dt}\right|$$

The definition of current is $I = dq/dt$. Consequently, the charge flow through the coil and onto the capacitor is given by

$$\frac{dq}{dt} = \frac{N}{R}\left|\frac{d\Phi}{dt}\right| \Rightarrow \frac{\Delta q}{\Delta t} = \frac{N}{R}\left|\frac{\Delta\Phi}{\Delta t}\right| \Rightarrow \Delta q = \frac{N}{R}\left|\Delta\Phi\right| = \frac{N}{R}(\Phi_{\text{f}} - \Phi_{\text{i}})$$

We are only interested in the total charge that flows due to the change in flux and not the details of the time dependence. In this case, the flux is changed by physically rotating the coil in the field. The flux is $\Phi = \vec{A}\cdot\vec{B} = AB\cos\theta$. The change in flux is

$$\Delta\Phi = AB(\cos\theta_f - \cos\theta_i) = \pi(0.020\text{ m})^2(55\times10^{-6}\text{ T})(\cos 30° - \cos 210°) = 1.2\times10^{-7}\text{ Wb}$$

Note that the field is 60° from the horizontal and the normal to the plane of the loop is vertical. The final angle, when $\vec{A}$ points down, is $\theta_f = 30°$, so the initial angle is $\theta_i = \theta_f + 180° = 210°$. The charge that flows onto the capacitor is

$$\Delta q = \frac{N}{R}(\Phi_f - \Phi_i) = \frac{200(1.2\times10^{-7}\text{ Wb})}{2.0\ \Omega} = 1.2\times10^{-5}\text{ C} \Rightarrow \Delta V_C = \frac{\Delta q}{C} = \frac{1.2\times10^{-5}\text{ C}}{1.0\times10^{-6}\text{ F}} = 12\text{ V}$$

34.67. Model: Assume we can ignore the sharp corners when the current changes abruptly.
Visualize: The changing current produces a changing flux, an induced emf, and a corresponding potential difference.
Solve: Break the current into time intervals over which the current is changing linearly or not at all. For the intervals 2 ms to 3 ms and 5 ms to 6 ms, the current does not change, so the potential difference is zero. For the interval 0 s to 2 ms, the current goes from 0 A to 2 A, so the potential difference is

$$\Delta V_L = -L\frac{dI}{dt} = -L\frac{\Delta I}{\Delta t} \Rightarrow \Delta V_L = -(10\times10^{-3}\text{ H})\frac{2\text{ A} - 0\text{ A}}{(2\text{ s} - 0\text{ s})\times10^{-3}} = -10\text{ V}$$

Similarly for the interval 3 ms to 5 ms, the potential difference is +20 V.

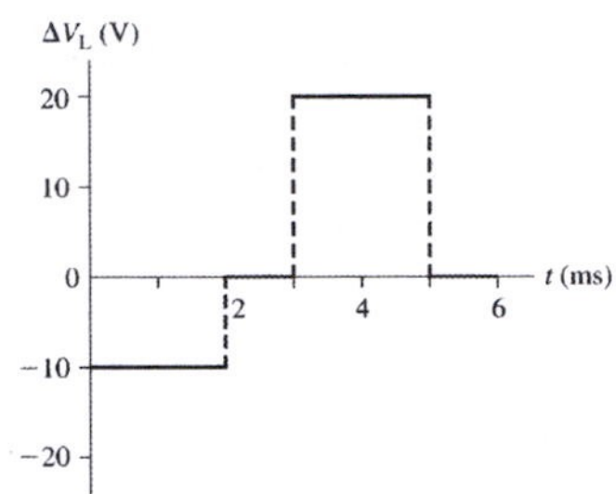

Assess: The potential difference is proportional to the negative slope of the current versus time graph.

34.71. Model: Assume any resistance is negligible.
Visualize: The potential difference across the inductor and capacitor oscillate.
Solve: **(a)** The current is $I(t) = I_0\cos\omega t = (0.50\text{ A})\cos\omega t$. Looking at Figure 34.46, we see that the capacitor is fully charged one-quarter cycle after the current is a maximum (or minimum), so the time needed is one quarter of a cycle. We have

$$\omega = \frac{1}{\sqrt{LC}} = \frac{1}{\sqrt{(20\times10^{-3}\text{ H})(8.0\times10^{-6}\text{ F})}} = 2.5\times10^3\text{ rad/s}$$

$$\Rightarrow T = \frac{2\pi}{\omega} = \frac{2\pi}{2.5\times10^3\text{ rad/s}} = 2.51\times10^{-3}\text{ s} = 2.51\text{ ms} \Rightarrow \Delta t = \tfrac{1}{4}T = 0.63\text{ ms}$$

(b) The maximum inductor current and maximum capacitor charge are related by $I_0 = \omega Q_0$. The potential across the capacitor is

$$\Delta V_C = \frac{Q_0}{C} = \frac{I_0}{\omega C} = \frac{0.50\text{ A}}{(2.5\times10^3\text{ rad/s})(8.0\times10^{-6}\text{ F})} = 25\text{ V}$$

34.73. Model: Assume any resistance is negligible.
Visualize: Energy in the capacitor and inductor oscillates as the charge and current oscillate, but the total energy is conserved.
Solve: The current through the inductor is zero when the charge on the capacitor is maximum. Thus the total energy is stored in the capacitor:

$$U_{\text{total}} = U_C = \frac{1}{2}\frac{Q_0}{C^2}$$

At a later time, when the capacitor's energy equals the inductor's energy, they each have half the total energy. Thus

$$U_C = U_L = \frac{1}{2}U_{\text{total}} \Rightarrow \frac{1}{2}\frac{Q}{C^2} = \frac{1}{2}\left(\frac{1}{2}\frac{Q_0}{C^2}\right) \Rightarrow Q = \frac{Q_0}{\sqrt{2}} = 0.707Q_0$$

34.79. Model: Assume an ideal inductor and an ideal (resistanceless) battery.
Visualize: Please refer to Figure P34.79.
Solve: **(a)** Because the switch has been open a long time, no current is flowing the instant before the switch is closed. A basic property of an ideal inductor is that the current through it cannot change instantaneously. This is because the potential difference $\Delta V_L = -L(dI/dt)$ would become infinite for an instantaneous change of current, and that is not physically possible. Because the current through the inductor was zero before the switch was closed, it must still be zero (or very close to it) immediately after the switch is closed. Consequently, the inductor has no effect on the circuit. It is simply a 10 Ω resistor and 20 Ω resistor in series with the battery. The equivalent resistance is 30 Ω, so the current through the circuit (including through the 20 Ω resistor) is $I = \Delta V_{\text{bat}}/R_{\text{eq}} = (30\text{ V})/(30\ \Omega) = 1.0\text{ A}$.
(b) After a long time, the currents in the circuit will reach steady values and no longer change. With steady currents, the potential difference across the inductor is $\Delta V_L = -L(dI/dt) = 0\text{ V}$. An ideal inductor has no resistance ($R = 0\ \Omega$), so the inductor simply acts like a wire. In this case, the inductor "shorts out" the 20 Ω resistor. All current from the 10 Ω resistor flows through the resistanceless inductor, so the current through the 20 Ω resistor is 0 A.
(c) When the switch has been closed a long time, and the inductor is shorting out the 20 Ω resistor, the current passing through the 10 Ω resistor and through the inductor is $I = (30\text{ V})/(10\ \Omega) = 3.0\text{ A}$. Because the current through an inductor cannot change instantaneously, the current must remain 3.0 A immediately after the switch reopens. This current must go somewhere (conservation of current), but now the open switch prevents the current from going back to the battery. Instead, it must flow upward through the 20 Ω resistor. That is, the current flows *around* the *LR* circuit consisting of the 20 Ω resistor and the inductor. This current will decay with time, with time constant $\tau = L/R$, but immediately after the switch reopens the current is 3.0 A.

35

Electromagnetic Fields and Waves

Exercises and Problems

35.1. Model: Apply the Galilean transformation of velocity.
Solve: **(a)** In the laboratory frame S, the speed of the proton is

$$v = \sqrt{(1.41\times10^6 \text{ m/s})^2 + (1.41\times10^6 \text{ m/s})^2} = 2.0\times10^6 \text{ m/s}$$

The angle the velocity vector makes with the positive y-axis is

$$\theta = \tan^{-1}\left(\frac{1.41\times10^6 \text{ m/s}}{1.41\times10^6 \text{ m/s}}\right) = 45°$$

(b) In the rocket frame S′, we need to first determine the vector $\vec{v}\,'$. Equation 34.1 yields:

$$\vec{v}\,' = \vec{v} - \vec{V} = (1.41\times10^6\hat{i} + 1.41\times10^6\hat{j}) \text{ m/s} - (1.00\times10^6\hat{i}) \text{ m/s} = (0.41\times10^6\hat{i} + 1.41\times10^6\hat{j}) \text{ m/s}$$

The speed of the proton is

$$v' = \sqrt{(0.41\times10^6 \text{ m/s})^2 + (1.41\times10^6 \text{ m/s})^2} = 1.47\times10^6 \text{ m/s}$$

The angle the velocity vector makes with the positive y'-axis is

$$\theta' = \tan^{-1}\left(\frac{0.41\times10^6 \text{ m/s}}{1.41\times10^6 \text{ m/s}}\right) = 16.2°$$

35.5. Model: Use the Galilean transformation of fields.
Visualize: Please refer to Figure EX35.5. We are given $\vec{V} = 1.0\times10^6\hat{i}$ m/s, $\vec{B} = 0.50\hat{k}$ T, and $\vec{E} = \left(\frac{1}{\sqrt{2}}\hat{i} + \frac{1}{\sqrt{2}}\hat{j}\right)\times10^6$ V/m.

Solve: Equation 35.11 gives the Galilean transformation equation for the electric field in the S and S′ frames: $\vec{E}' = \vec{E} + \vec{V}\times\vec{B}$. The electric field from the moving rocket is

$$\vec{E}' = (\hat{i} + \hat{j})0.707\times10^6 \text{ V/m} + (1.0\times10^6\hat{i} \text{ m/s})\times(0.50\hat{k} \text{ T}) = (0.707\times10^6\hat{i} + 0.207\times10^6\hat{j}) \text{ V/m}$$

$$\theta = \tan^{-1}\left(\frac{0.207\times10^6 \text{ V/m}}{0.707\times10^6 \text{ V/m}}\right) = 16.3° \text{ above the } x'\text{-axis}$$

35.9. Model: The displacement current is numerically equal to the current in the wires leading to and from the capacitor.
Solve: During the charging process, a parallel-plate capacitor develops charge as a function of time. If the charge on a capacitor plate is Q at time t, then $Q = CV_C$, where V_C is the voltage across the capacitor plates. Taking the derivative,

$$\frac{dQ}{dt} = I = C\frac{dV_C}{dt} \Rightarrow \frac{dV_C}{dt} = \frac{I}{C} = \frac{1.0 \text{ A}}{1.0\ \mu\text{F}} = 1.0\times10^6 \text{ V/s}$$

35.11. Model: The displacement current is numerically equal to the current in the wires leading to and from the capacitor.
Solve: The process of charging increases the charge on the plates of a parallel-plate capacitor. The charge Q on a capacitor plate at time t is $Q = CV_C$, where V_C is the voltage across the capacitor plates. Taking the derivative,

$$I_{\text{disp}} = I = \frac{dQ}{dt} = C\frac{dV_C}{dt} = \frac{\varepsilon_0 A}{d}\frac{dV_C}{dt} = \frac{(8.85\times10^{-12}\ \text{C}^2/\text{N m}^2)(0.050\ \text{m})^2}{0.50\times10^{-3}\ \text{m}}(500{,}000\ \text{V/s}) = 22\ \mu\text{A}$$

35.15. Model: Electromagnetic waves are sinusoidal.
Solve: **(a)** The electric field is $E_y = E_0 \cos(kx - \omega t)$, where $E_0 = 20.0$ V/m and $k = 6.28 \times 10^8\ \text{m}^{-1}$. The wavelength is

$$\lambda = \frac{2\pi}{k} = \frac{2\pi}{6.28\times10^8\ \text{m}^{-1}} = 1.00\times10^{-8}\ \text{m} = 10.0\ \text{nm}$$

(b) The frequency is

$$f = \frac{c}{\lambda} = \frac{3.0\times10^8\ \text{m/s}}{1.00\times10^{-8}\ \text{m}} = 3.00\times10^{16}\ \text{Hz}$$

(c) The magnetic field amplitude is

$$B_0 = \frac{E_0}{v_{\text{em}}} = \frac{20.0\ \text{V/m}}{3.0\times10^8\ \text{m/s}} = 6.67\times10^{-8}\ \text{T}$$

35.21. Model: A radio wave is an electromagnetic wave.
Solve: **(a)** The energy transported per second by the radio wave is 25 kW, or 25×10^3 J/s. This energy is carried uniformly in all directions. From Equation 35.37, the light intensity is

$$I = \frac{P}{A} = \frac{P}{4\pi r^2} = \frac{25\times10^3\ \text{W}}{4\pi(30\times10^3\ \text{m})^2} = 2.2\times10^{-6}\ \text{W/m}^2$$

(b) Using Equation 35.37 again,

$$I = \frac{c\varepsilon_0}{2}E_0^2 \Rightarrow 2.2\times10^{-6}\ \text{W/m}^2 = \frac{(3\times10^8\ \text{m/s})(8.85\times10^{-12}\ \text{C}^2/\text{N m}^2)}{2}E_0^2 \Rightarrow E_0 = 0.041\ \text{V/m}$$

35.27. Model: Use Malus's law for polarized light.
Visualize:

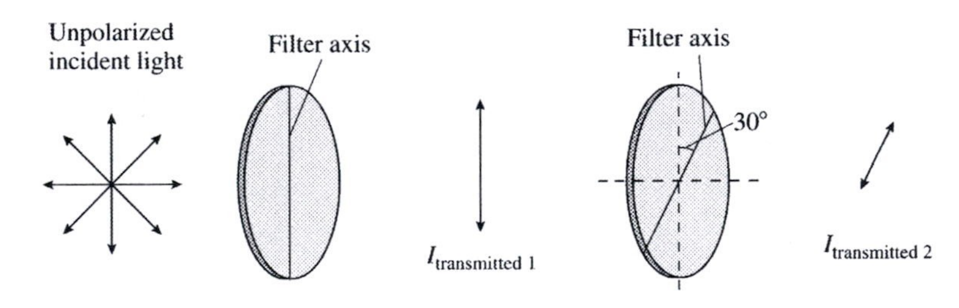

Solve: For unpolarized light, the electric field vector varies randomly through all possible values of θ. Because the *average* value of $\cos^2\theta$ is $\frac{1}{2}$, the intensity transmitted by a polarizing filter is $I_{\text{transmitted}} = \frac{1}{2}I_0$. On the other hand, for polarized light $I_{\text{transmitted}} = I_0 \cos^2\theta$. Therefore,

$$I_{\text{transmitted 2}} = I_{\text{transmitted 1}}\cos^2\theta = \tfrac{1}{2}I_0\cos^2\theta = \tfrac{1}{2}(350\ \text{W/m}^2)\cos^2 30° = 131\ \text{W/m}^2$$

Assess: Note that any particular *wave* has a clear polarization. It is only in a "sea" of waves that the resultant wave has no polarization.

35.29. Model: Assume the electric and magnetic fields are uniform.
Visualize: Please refer to Figure P35.29.
Solve: The force on the proton, which is the sum of the electric and magnetic forces, is

$$\vec{F} = \vec{F}_E + \vec{F}_B = -F\cos 30°\hat{i} + F\sin 30°\hat{j} = (-2.77\hat{i} + 1.60\hat{j})\times 10^{-13}\text{ N}$$

Since $\vec{v}$ points out of the page, the magnetic force is $\vec{F}_B = e\vec{v}\times\vec{B} = 1.60\times 10^{-13}\,\hat{j}$ N. Thus

$$\vec{F}_E = e\vec{E} = \vec{F} - \vec{F}_B = -2.77\times 10^{-13}\,\hat{i}\text{ N} \Rightarrow \vec{E} = \vec{F}_E/e = -1.73\times 10^{6}\,\hat{i}\text{ V/m}$$

That is, the electric field is $\vec{E}$ = (1.73×10^6 V/m, left).

35.33. Model: Use the Galilean transformation of fields.
Visualize:

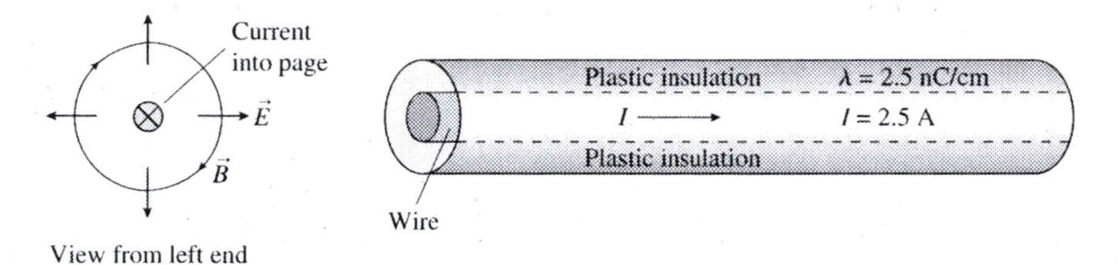

A current of 2.5 A flows to the right through the wire, and the plastic insulation has a charge of linear density λ = 2.5 n C/cm.
Solve: The magnetic field B at a distance r from the wire is

$$\vec{B} = \left(\frac{\mu_0 I}{2\pi r},\text{ clockwise seen from left}\right)$$

On the other hand, $\vec{E}$ is radially out along $\hat{r}$, that is,

$$\vec{E} = \frac{\lambda}{2\pi\varepsilon_0 r}\hat{r}$$

As the mosquito is 1.0 cm from the center of the wire at the top of the wire,

$$B = \frac{(4\pi\times 10^{-7}\text{ T m/A})(2.5\text{ A})}{2\pi(0.010\text{ m})} = 5.0\times 10^{-5}\text{ T}$$

$$E = \frac{(2.5\times 10^{-7}\text{ C/m})(2)(9.0\times 10^{9}\text{ N m}^2/\text{C}^2)}{0.010\text{ m}} = 4.5\times 10^{5}\ \frac{\text{V}}{\text{m}}$$

where the direction of $\vec{B}$ is out of the page and the direction of $\vec{E}$ is radially outward. In the mosquito's frame (let us call it S′), we want $\vec{B}' = \vec{0}$ T. Thus,

$$\vec{B}' = \vec{B} - \frac{1}{c^2}\vec{V}\times\vec{E} = 0 \Rightarrow \vec{B} = \frac{1}{c^2}\vec{V}\times\vec{E}$$

Because $\vec{V}\times\vec{E}$ must be in the direction of $\vec{B}$ and $\vec{E}$ is radially outward, according to the right-hand rule $\vec{V}$ must be along the direction of the current. The magnitude of the velocity is

$$V = \frac{c^2 B}{E} = \frac{(3.0\times 10^{8}\text{ m/s})^2(5.0\times 10^{-5}\text{ T})}{4.5\times 10^{5}\text{ V/m}} = 1.0\times 10^{7}\text{ m/s}$$

The mosquito must fly at 1.0 ×10^7 m/s parallel to the current. This is highly unlikely to happen unless the mosquito is from Planet Krypton, like Superman.

35.41. Model: $\vec{E}$ and $\vec{B}$ are perpendicular in electromagnetic waves, and their magnitudes are related.
Solve: **(a)** Since $\vec{E} \perp \vec{B}$, the dot product must be zero.

$$\vec{E} \cdot \vec{B} = 0 = (200)(7.3) + (300)(-7.3) + (-50)a$$
$$\Rightarrow a = -14.6$$

Since B_0 multiplies the complete vector $\vec{B}$ it does not effect the calculation for a. Requiring $|\vec{E}| = c|\vec{B}|$ and squaring yields

$$\left((200)^2 + (300)^2 + (-50)^2\right)(\text{V/m})^2 = B_0^2\left(3.0\times10^8 \text{ m/s}\right)^2\left((7.3)^2 + (-7.3)^2 + (-14.6)^2\right)(\mu\text{T})^2$$
$$\Rightarrow B_0 = 6.8\times10^{-2}$$

(b) The Poynting vector is

$$\vec{S} = \mu_0^{-1}\vec{E}\times\vec{B}$$
$$= \mu_0^{-1}B_0\left(10^{-6}\right)\begin{Bmatrix}\left[(300)(-14.6)-(-7.3)(-50)\right]\hat{i} \\ +\left[(-50)(7.3)-(-14.6)(200)\right]\hat{j} \\ +\left[(200)(-7.3)-(7.3)(300)\right]\hat{k}\end{Bmatrix}$$
$$= \mu_0^{-1}B_0\left(10^{-3}\right)\left[-4.75\hat{i} + 2.56\hat{j} - 3.65\hat{k}\right] = -260\hat{i} + 140\hat{j} - 200\hat{k} \text{ W/m}^2$$

Assess: A quick check yields $\vec{E}\cdot\vec{S} = 0$ and $\vec{B}\cdot\vec{S} = 0$.

35.49. Model: The laser beam is an electromagnetic wave.
Solve: The maximum intensity of the laser beam is determined by the maximum electric field strength in air. Thus the maximum power delivered by the beam is

$$P = IA = \frac{c\varepsilon_0}{2}E_0^2 A$$
$$= \frac{\left(3\times10^8 \text{ m/s}\right)\left(8.85\times10^{-12} \text{ C}^2/\text{Nm}^2\right)}{2}\left(3.0\times10^6 \text{ V/m}\right)^2 \pi\left(0.050 \text{ m}\right)^2$$
$$= 9.4\times10^7 \text{ W}$$

35.53. Model: Assume that the black paper absorbs the light completely. Use the particle model for the paper.
Visualize:

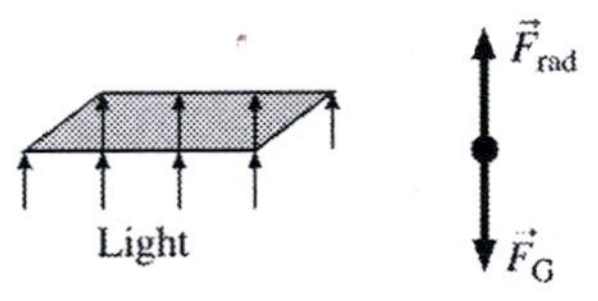

For the black paper to be suspended, the radiation-pressure force must be equal to the gravitational force on the paper.
Solve: From Equation 35.39, $F_{\text{rad}} = p_{\text{rad}}A = IA/c$. Hence,

$$I = \frac{c}{A}F_{\text{rad}} = \frac{c}{A}F_{\text{G}} = \frac{\left(3.0\times10^8 \text{ m/s}\right)\left(1.0\times10^{-3} \text{ kg}\right)\left(9.8 \text{ m/s}^2\right)}{\left(8.5 \text{ inch}\times11 \text{ inch}\right)\left(2.54\times10^{-2} \text{ m/inch}\right)^2} = 4.9\times10^7 \text{ W/m}^2$$

35.55. Model: Use the particle model for the astronaut.
Solve: According to Newton's third law, the force of the radiation on the astronaut is equal to the momentum delivered by the radiation. For this force we have

$$F = p_{\text{rad}}A = \frac{P}{c} = \frac{1000 \text{ W}}{3.0\times10^8 \text{ m/s}} = 3.333\times10^{-6} \text{ N}$$

Using Newton's second law, the acceleration of the astronaut is

$$a = \frac{3.333\times10^{-6} \text{ N}}{80 \text{ kg}} = 4.167\times10^{-8} \text{ m/s}^2$$

Using $v_f = v_i + a(t_f - t_i)$ and a time equal to the lifetime of the batteries,

$$v_f = 0 \text{ m/s} + (4.167 \times 10^{-8} \text{ m/s}^2)(3600 \text{ s}) = 1.500 \times 10^{-4} \text{ m/s}$$

The distance traveled in the first hour is calculated as follows:

$$v_f^2 - v_i^2 = 2a(\Delta s)_{\text{first hour}}$$

$$\Rightarrow (1.500 \times 10^{-4} \text{ m/s})^2 - (0 \text{ m/s})^2 = 2(4.167 \times 10^{-8} \text{ m/s}^2)(\Delta s)_{\text{first hour}} \Rightarrow (\Delta s)_{\text{first hour}} = 0.270 \text{ m}$$

This means the astronaut must cover a distance of 5.0 m − 0.27 m = 4.73 m in a time of 9 hours. The acceleration is zero during this time. The time it will take the astronaut to reach the space capsule is

$$\Delta t = \frac{4.73 \text{ m}}{1.500 \times 10^{-4} \text{ m/s}} = 31{,}533 \text{ s} = 8.76 \text{ hours}$$

Because this time is less than 10 hours, the astronaut is able to make it safely to the space capsule.

AC CIRCUITS

36

Exercises and Problems

36.3. Model: A phasor is a vector that rotates counterclockwise around the origin at angular velocity ω.
Solve: The emf is

$$\mathcal{E} = \mathcal{E}_0 \cos\omega t = (50\text{ V})\cos(2\pi \times 110\text{ rad/s} \times 3.0 \times 10^{-3}\text{ s}) = (50\text{ V})\cos(2.074\text{ rad}) = (50\text{ V})\cos 119°$$

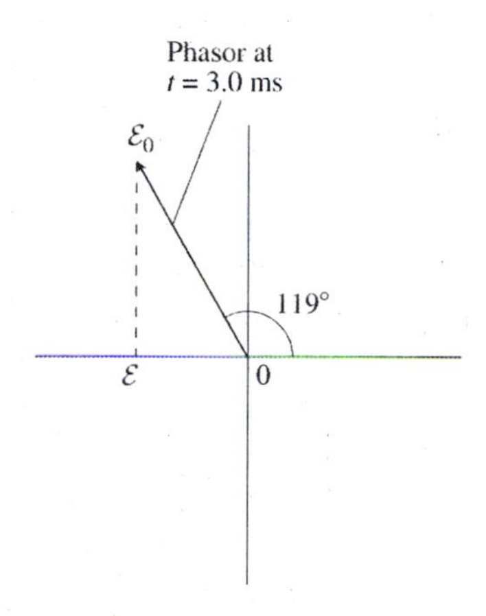

36.7. Visualize: Figure EX36.7 shows a simple one-capacitor circuit.
Solve: **(a)** The capacitive reactance at $\omega = 2\pi f = 2\pi(100\text{ Hz}) = 628.3$ rad/s is

$$X_C = \frac{1}{\omega C} = \frac{1}{(628.3\text{ rad/s})(0.30 \times 10^{-6}\text{ F})} = 5305\ \Omega$$

$$\Rightarrow I_C = \frac{V_C}{X_C} = \frac{10\text{ V}}{5.305 \times 10^3\ \Omega} = 1.88 \times 10^{-3}\text{ A} = 1.88\text{ mA}$$

(b) The capacitive reactance at $\omega = 2\pi(100\text{ kHz}) = 628{,}300$ rad/s is

$$X_C = \frac{1}{\omega C} = \frac{1}{(6.283 \times 10^5\text{ rad/s})(0.30 \times 10^{-6}\text{ F})} = 5.305\ \Omega$$

$$\Rightarrow I_C = \frac{V_C}{X_C} = \frac{10\text{ V}}{5.305\ \Omega} = 1.88\text{ A}$$

Assess: Using reactance is just like using resistance in Ohm's law. Because $X_C \propto \omega^{-1}$, X_C decreases with an increase in ω, as observed above.

36.15. Model: The current and voltage of a resistor are in phase, but the capacitor current leads the capacitor voltage by 90°.
Visualize: Please refer to Figure EX36.15.
Solve: From Equations 36.13 and 36.14, the peak voltages are $V_R = IR$ and $V_C = IX_C$, where

$$X_C = \frac{1}{\omega C} = \frac{1}{2\pi fC} = \frac{1}{2\pi(1\times10^4 \text{ Hz})(80\times10^{-4} \text{ F})} = 199\ \Omega$$

The peak current is

$$I = \frac{\mathcal{E}_0}{\sqrt{X_C^2 + R^2}} = \frac{10 \text{ V}}{\sqrt{(199\ \Omega)^2 + (150\ \Omega)^2}} = 0.0401 \text{ A}$$

Thus, $V_R = (0.0401 \text{ A})(150\ \Omega) = 6.0$ V and $V_C = IX_C = (0.0401 \text{ A})(199\ \Omega) = 8.0$ V.

36.21. Model: The AC current through an inductor lags the inductor voltage by 90°.
Solve: **(a)** From Equation 36.21,

$$I_L = 50 \text{ mA} = \frac{V_L}{X_L} = \frac{V_L}{2\pi fL} \Rightarrow f = \frac{5.0 \text{ V}}{2\pi\left(50\times10^{-3} \text{ A}\right)\left(500\times10^{-6} \text{ H}\right)} = 3.2\times10^4 \text{ Hz}$$

(b) The current and voltage for a simple one-inductor circuit are

$$i_L = I_L\cos\left(\omega t - \tfrac{1}{2}\pi\right) \qquad v_L = V_L\cos\omega t$$

For $i_L = I_L$, $\omega t - \frac{1}{2}\pi$ must be equal to $2n\pi$, where $n = 0, 1, 2, \ldots$. This means $\omega t = \left(2n\pi + \frac{1}{2}\pi\right)$. Thus, the instantaneous value of the emf at the instant when $i_L = I_L$ is $v_L = 0$ V.

36.27. Visualize: The circuit looks like Figure 36.17.
Solve: **(a)** The impedance of the circuit for a frequency of 3000 Hz is

$$Z = \sqrt{R^2 + \left(X_L - X_C\right)^2} = \sqrt{(50\ \Omega)^2 + \left[2\pi\left(3\times10^3 \text{ Hz}\right)\left(3.3\times10^{-3} \text{ H}\right) - \frac{1}{2\pi\left(3\times10^3 \text{ Hz}\right)\left(480\times10^{-9} \text{ F}\right)}\right]^2}$$

$$= \sqrt{(50\ \Omega)^2 + \left(62.20\ \Omega - 110.52\ \Omega\right)^2} = 69.53\ \Omega \approx 70\ \Omega$$

The peak current is

$$I = \frac{\mathcal{E}_0}{Z} = \frac{5.0 \text{ V}}{69.53\ \Omega} = 0.072 \text{ A} = 72 \text{ mA}$$

The phase angle is

$$\phi = \tan^{-1}\left[\frac{X_L - X_C}{R}\right] = \tan^{-1}\left(\frac{-48.32\ \Omega}{50\ \Omega}\right) = -44°$$

(b) For 4000 Hz, $Z = 50.0\ \Omega$ $I = 0.100$ A, and $\phi = 0°$.
(c) For 5000 Hz, $Z = 62.42\ \Omega \approx 62\ \Omega$, $I = 0.080$ A, and $\phi = 37°$
The following table summarizes the results.

	f = 3000 Hz	f = 4000 Hz	f = 5000 Hz
Z (Ω)	70	50	62
I (A)	0.072	0.100	0.080
ϕ	–44°	0°	37°

36.33. Model: The energy supplied by the emf source to the RLC circuit is dissipated by the resistor. Because of the phase difference between the current and the emf, the energy dissipated is $P_R = I_{rms}\mathcal{E}_{rms}\cos\phi = I_{rms}V_{rms}$ (Equation 36.47).
Solve: From Equation 36.28,

$$V_R = \mathcal{E}_0\cos\phi \Rightarrow V_{rms} = \mathcal{E}_{rms}\cos\phi$$

Using Ohm's law,

$$R = \frac{V_{\text{rms}}}{I_{\text{rms}}} = \frac{\mathcal{E}_{\text{rms}}\cos\phi}{I_{\text{rms}}} = \frac{(120\text{ V})(0.87)}{2.4\text{ A}} = 44\ \Omega$$

36.35. Solve: (a) From Equation 36.14,

$$V_{\text{R}} = \frac{\mathcal{E}_0 R}{\sqrt{R^2 + (\omega_{\text{cap}} C)^{-2}}} = \frac{\mathcal{E}_0}{2} \Rightarrow R^2 + \frac{1}{\omega_{\text{res}}^2 C^2} = 4R^2 \Rightarrow \omega_{\text{res}} = \frac{1}{\sqrt{3}RC}$$

(b) At this frequency,

$$V_{\text{C}} = IX_{\text{C}} = \frac{V_{\text{R}}}{R}\left(\frac{1}{\omega_{\text{res}}C}\right) = \frac{(\mathcal{E}_0/2)}{R}\left(\sqrt{3}RC\right)\frac{1}{C} = \frac{\sqrt{3}}{2}\mathcal{E}_0$$

(c) The crossover frequency is

$$\omega_{\text{c}} = \frac{1}{RC} = 6280\text{ rad/s} \Rightarrow \omega_{\text{res}} = \omega_{\text{c}}/\sqrt{3} = 3630\text{ rad/s}$$

36.37. Visualize: Please refer to Figure P36.37.
Solve: (a) The voltage across the capacitor is

$$V_{\text{C}} = IX_{\text{C}} = \frac{\mathcal{E}_0}{\sqrt{R^2 + X_{\text{C}}^2}}X_{\text{C}} = \frac{\mathcal{E}_0(1/\omega C)}{\sqrt{R^2 + (1/\omega C)^2}} = \frac{\mathcal{E}_0}{\sqrt{(\omega RC)^2 + 1}}$$

$$= \frac{10\text{ V}}{\sqrt{4\pi^2 f^2 (16\ \Omega)^2 (1.0\times10^{-6}\text{ F})^2 + 1}} = \frac{10\text{ V}}{\sqrt{1 + (1.0106\times10^{-8}\text{ s}^2) f^2}}$$

The values of V_{C} at a few frequencies are in the following table.

f (kHz)	V_{C} (V)
1	9.95
3	9.57
10	7.05
30	3.15
100	0.990

(b)

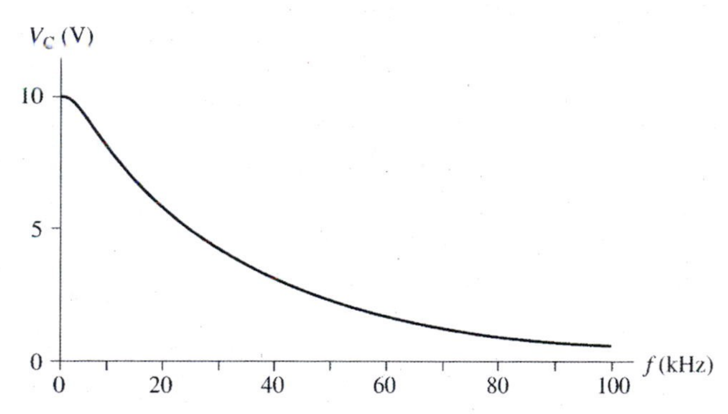

Assess: For the voltage across the capacitor, the circuit is a low-pass filter.

36.43. Solve: From Equation 32.32, the voltage of a discharging capacitor is

$$V_{\text{C}} = V_0 e^{-t/RC} \Rightarrow \tfrac{1}{2}V_0 = V_0 e^{-(2.5\text{ ms})/RC} \Rightarrow \ln\left(\tfrac{1}{2}\right) = -\frac{(2.5\text{ ms})}{RC} \Rightarrow RC = \frac{-(2.5\times10^{-3}\text{ s})}{\ln 0.5} = 3.61\times10^{-3}\text{ s}$$

The crossover frequency for a low-pass circuit is

$$\omega_{\text{c}} = \frac{1}{RC} \Rightarrow f_{\text{c}} = \frac{1}{2\pi RC} = \frac{1}{2\pi(3.61\times10^{-3}\text{ s})} = 44\text{ Hz}$$

36.47. Model: While the AC current through an inductor lags the inductor voltage by 90°, the current and the voltage are in phase for a resistor.
Visualize: Series elements have the same current, so we start with a common current phasor I for the inductor and resistor,

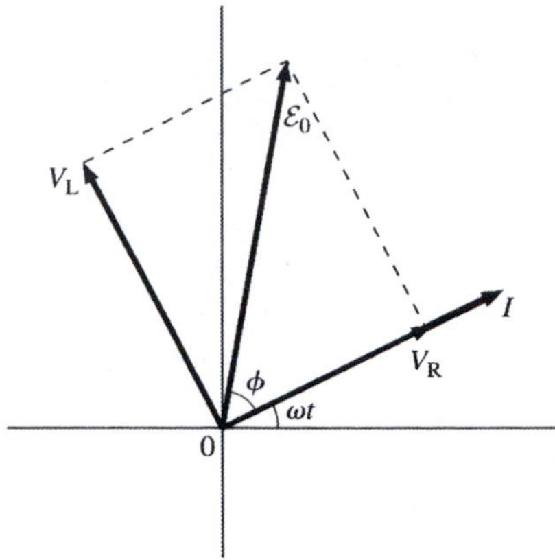

Please refer to Figure P36.47.
Solve: Because we have a series RL circuit, the current through the resistor and the inductor is the same. The voltage phasor V_R is along the same direction as the current phasor I. The voltage phasor V_L is ahead of the current phasor by 90°.
(a) From the phasors in the figure, $\mathcal{E}_0 = \sqrt{V_L^2 + V_R^2}$. Noting that $V_L = I\omega L$ and $V_R = IR$, $\mathcal{E}_0 = I\sqrt{\omega^2 L^2 + R^2}$ and

$$I = \frac{\mathcal{E}_0}{\sqrt{R^2 + \omega^2 L^2}} \qquad V_R = \frac{\mathcal{E}_0 R}{\sqrt{R^2 + \omega^2 L^2}} \qquad V_L = \frac{\mathcal{E}_0 \omega L}{\sqrt{R^2 + \omega^2 L^2}}$$

(b) As $\omega \to 0$ rad/s, $V_R \to \mathcal{E}_0 R/R = \mathcal{E}_0$ and as $\omega \to \infty$, $V_R \to 0$ V.
(c) The LR circuit will be a low-pass filter, if the output is taken from the resistor. This is because V_R is maximum when ω is low and goes to zero when ω becomes large.
(d) At the cross-over frequency, $V_L = V_R$. Hence,

$$I\omega_c L = IR \Rightarrow \omega_c = \frac{R}{L}$$

36.49. Visualize: The circuit looks like the one in Figure 36.17.
Solve: **(a)** The impedance of the circuit is

$$Z = \sqrt{R^2 + (X_L - X_C)^2} = \sqrt{(100\ \Omega)^2 + \left(2\pi(60\text{ Hz})(0.15\text{ H}) - \frac{1}{2\pi(60\text{ Hz})(30\times10^{-6}\text{ F})}\right)^2}$$
$$= \sqrt{(100\ \Omega)^2 + (56.55\ \Omega\ -\ 88.42\ \Omega)^2} = 104.96\ \Omega.$$

The peak current is

$$I = \frac{\mathcal{E}_0}{Z} = \frac{(120\text{ V})\sqrt{2}}{104.96} = 1.617\text{ A} = 1.62\text{ A}$$

(b) The phase angle is

$$\phi = \tan^{-1}\left(\frac{X_L - X_C}{R}\right) = \tan^{-1}\left(\frac{56.55\ \Omega\ -\ 88.42\ \Omega}{100\ \Omega}\right) = -17.7°$$

Because ϕ is negative, this is a capacitive circuit.
(c) The average power loss is

$$P_R = I_{rms}\mathcal{E}_{rms}\cos\phi = \frac{\mathcal{E}_{rms}^2}{R}\cos\phi = \frac{(120\text{ V})^2}{100\ \Omega}\cos(-17.7°) = 137\text{ W}$$

36.53. Model: An RLC is driven above the resonance frequency when the circuit current lags the emf.
Visualize: The circuit looks like the one in Figure 36.17.
Solve: The phase angle is +30° since the circuit is above resonance, and $X_L > X_C$. Thus

$$\tan 30° = \frac{X_L - X_C}{R} \Rightarrow X_L - X_C = R\tan 30°$$

The impedance of the circuit is

$$Z = \sqrt{R^2 + (X_L - X_C)^2} = \sqrt{R^2 + (R\tan 30°)^2} = R\sqrt{1 + (\tan 30°)^2}$$

The peak current is

$$I = \frac{\mathcal{E}_0}{Z} = \frac{10\text{ V}}{(50\Omega)\sqrt{1 + (\tan 30°)^2}} = 0.173\text{ A}$$

Assess: Remember to put your calculator back into degrees mode when calculating tan 30°.

36.55. Visualize: The circuit looks like the one in Figure 36.17.
Solve: (a) The instantaneous current is $i = I\cos(\omega t - \phi)$. The phase angle is

$$\phi = \tan^{-1}\left[\frac{X_L - X_C}{R}\right] = \tan^{-1}\left(\frac{-48.32\ \Omega}{50\ \Omega}\right) = -44°$$

Since $i = I\cos(\omega t - \phi)$, $i = I$ implies that $\omega t - \phi = 0$ rad. That is, $\omega t = \phi = -44°$. Thus,

$$\mathcal{E} = \mathcal{E}_0 \cos\omega t = (5.0\text{ V})\cos(-44°) = 3.6\text{ V}$$

(b) $i = 0$ A and is decreasing implies that $\omega t - \phi = \frac{1}{2}\pi$ rad. That is, $\omega t = \frac{1}{2}\pi + \phi = 90° - 44° = 46°$. Thus, $\mathcal{E} = (5.0\text{ V})\cos(46°) = 3.5\text{ V}$.
(c) $i = -I$ implies $\omega t - \phi = \pi$ rad. That is, $\omega t = \pi + \phi = 180° - 44° = 136°$. Thus, $\mathcal{E} = (5.0\text{ V})\cos 136° = -3.6\text{ V}$.

36.61. Model: The filament in a light bulb acts as a resistor.
Visualize: Please refer to Figure P36.61.
Solve: A bulb labeled 40 W is designed to dissipate an average of 40 W of power at a voltage of V_{rms} = 120 V. From Equation 36.39, the resistance of the 40 W light bulb is

$$R_{40} = \frac{V_{rms}^2}{P_{40}} = \frac{(120\text{ V})^2}{40\text{ W}} = 360\ \Omega$$

Likewise, $R_{60} = 240\ \Omega$ and $R_{100} = 144\ \Omega$. The rms current through the 40 W and 60 W bulbs, which are in series, is

$$I_{60} = I_{40} = \frac{\mathcal{E}_0}{600\ \Omega} = \frac{120\text{ V}}{600\ \Omega} = 0.20\text{ A}$$

The rms voltage across the light bulbs in series are

$$V_{40} = I_{40}R_{40} = (0.20\text{ A})(360\ \Omega) = 72 \qquad V_{60} = I_{60}R_{60} = (0.20\text{ A})(240\ \Omega) = 48\text{ V}$$

The voltage across the 100 W light bulb is V_{100} = 120 V. The powers dissipated in the light bulbs in series are

$$P_{40} = V_{40}I_{40} = (72\text{ V})(0.20\text{ A}) = 14.4\text{ W} \qquad P_{60} = V_{60}I_{60} = (48\text{ V})(0.20\text{ A}) = 9.6\text{ W}$$

The power dissipated in the 100 W bulb is

$$P_{100} = \frac{V_{100}^2}{R_{100}} = \frac{(120\text{ V})^2}{144\ \Omega} = 100\text{ W}$$

36.63. Model: Assume the motor is a series RLC circuit.

Solve: **(a)** From Equation 36.44, the average power is $P_{\text{source}} = I_{\text{rms}}\mathcal{E}_{\text{rms}}\cos\phi$. Thus, the power factor is

$$\cos\phi = \frac{P_{\text{source}}}{I_{\text{rms}}\mathcal{E}_{\text{rms}}} = \frac{800\text{ W}}{(120\text{ V})(8.0\text{ A})} = 0.833$$

(b) The resistance voltage is $V_{\text{rms}} = \mathcal{E}_{\text{rms}}\cos\phi = (120\text{ V})(0.833) = 100\text{ V}$.

(c) Using Ohm's law, the resistance of the motor is

$$R = \frac{V_{\text{rms}}}{I_{\text{rms}}} = \frac{100\text{ V}}{8.0\text{ A}} = 12.5\ \Omega$$

(d) From the results of part (a) and from Equation 36.27, the phase angle is

$$\phi = \cos^{-1}(0.833) = 33.6° = \tan^{-1}\left(\frac{X_{\text{L}} - X_{\text{C}}}{R}\right)$$

$$\Rightarrow X_{\text{L}} - X_{\text{C}} = (12.5\ \Omega)\tan 33.6° = 8.292\ \Omega$$

The phase angle becomes zero and the power factor becomes 1.0 by increasing the capacitive reactance by 8.292 Ω. The capacitor with this reactance is

$$C = \frac{1}{2\pi(60\text{ Hz})(8.292\ \Omega)} = 3.20\times10^{-4}\text{ F} = 320\ \mu\text{F}$$

37

RELATIVITY

Exercises and Problems

37.3. Model: S is the ground's frame of reference and S′ is the sprinter's frame of reference. Frame S′ moves relative to frame S with speed v.

Visualize:

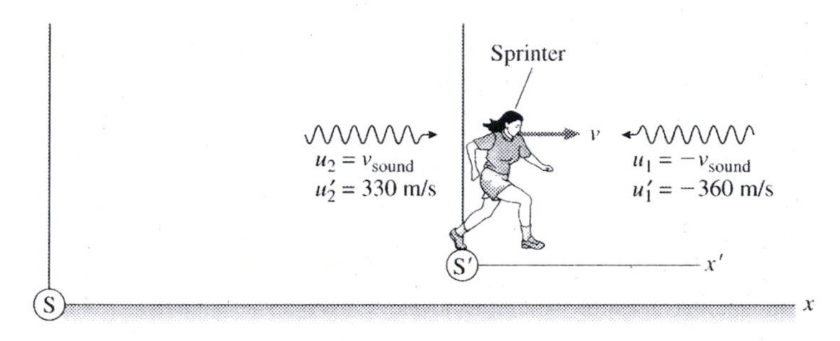

Solve: The speed of a sound wave is measured relative to its medium. The medium is still air on the ground, which is our frame S. The sprinter travels to the right with reference frame S′ at velocity v. Using the Galilean transformations of velocity,

$$u_1' = -360 \text{ m/s} = u_1 - v = -v_{\text{sound}} - v \qquad u_2' = 330 \text{ m/s} = u_2 - v = v_{\text{sound}} - v$$

Adding the two above equations,

$$-30 \text{ m/s} = -2v \Rightarrow v_{\text{sprinter}} = 15 \text{ m/s}$$

From the first equation,

$$-360 \text{ m/s} = -v_{\text{sound}} - (15 \text{ m/s}) \Rightarrow v_{\text{sound}} = 345 \text{ m/s}$$

Assess: Notice that the Galilean transformations use velocities and not speeds. It is for that reason $u_1' = -360$ m/s.

37.11. Model: Bianca and firecrackers 1 and 2 are in the same reference frame. Light from both firecrackers travels toward Bianca at 300 m/μs.

Visualize:

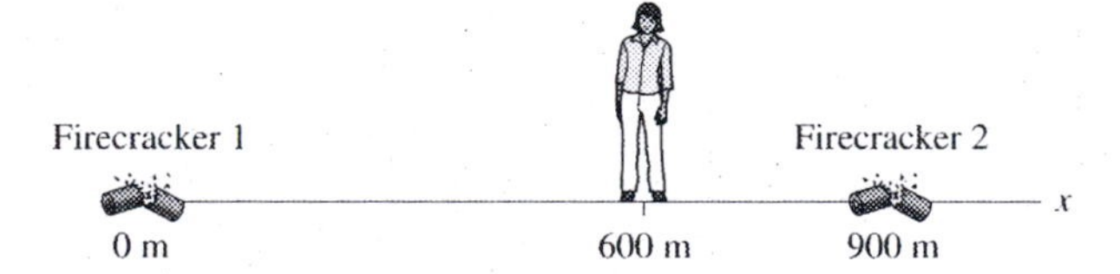

Solve: The flash from firecracker 1 takes 2.0 μs to reach Bianca $(600 \text{ m} \div 300 \text{ m}/\mu\text{s})$. The firecracker exploded at t_1 = 1.0 μs because it reached Bianca's eye at 3.0 μs. The flash from the firecracker 2 takes 1.0 μs to reach Bianca. Since firecrackers 1 and 2 exploded simultaneously, the explosion occurs at t_2 = 1.0 μs. So, the light from firecracker 2 reaches Bianca's eye at 2.0 μs. Although the events are simultaneous, Bianca *sees* them occurring at different times.

37.13. Model: You and your assistant are in the same reference frame. Light from the two lightning bolts travels toward you and your assistant at 300 m/μs. You and your assistant have synchronized clocks.
Visualize:

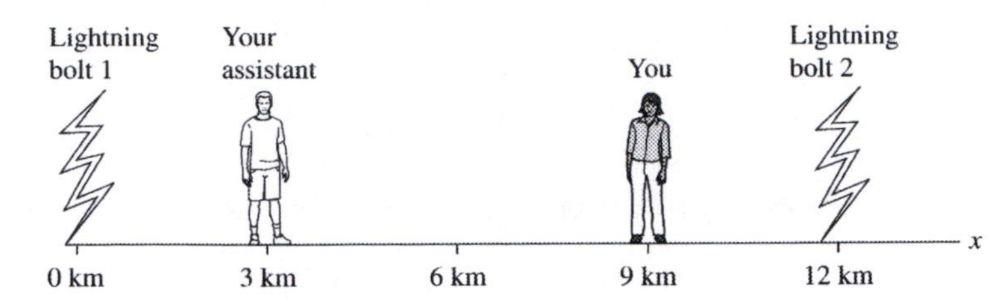

Solve: Bolt 1 hits 9.0 km away, so the light takes 30 μs to reach you (9000 m ÷ 300 m/μs). You see this flash at t = 50 μs, so the lightning hit at t_1 = 20 μs. Light from bolt 2, which hits 3.0 km away, takes 10 μs to reach you. You see it at 10 μs, so the lightning hit at t_2 = 0 μs. The strikes are not simultaneous. Bolt 2 hits first, 20 μs before bolt 1. Your assistant is in your inertial reference frame, so your assistant agrees that bolt 2 hits first, 20 μs before bolt 1.
Assess: A simple calculation would show that your assistant *sees* the flashes at the same time. When the flashes are seen is not the same as when the events happened.

37.17. Model: Let the moving clock be in frame S′ and an identical at-rest clock be in frame S.
Solve: The ticks being measured are those of the moving clock. The interval between 2 ticks is measured by the same clock in S′—namely, the clock that is ticking—so this is the proper time: $\Delta t' = \Delta\tau$. The rest clock measures a longer interval Δt between two ticks of the moving clock. These are related by

$$\Delta\tau = \sqrt{1-\beta^2}\;\Delta t$$

The moving clock ticks at half the rate of the rest clock when $\Delta\tau = \frac{1}{2}\Delta t$. Thus

$$\sqrt{1-\beta^2} = \sqrt{1-v^2/c^2} = 1/2 \Rightarrow v = c\sqrt{1-(1/2)^2} = 0.866c$$

37.33. Model: The earth and the other galaxy are inertial reference frames. Let the earth be frame S and the other galaxy be frame S′. S′ moves with $v = +0.2c$. The quasar's speed in frame S is $u = +0.8c$.
Solve: Using the Lorentz velocity transformation equation,

$$u' = \frac{u-v}{1-uv/c^2} = \frac{0.8c-0.2c}{1-(0.8c)(0.2c)/c^2} = 0.71c$$

Assess: In Newtonian mechanics, the Galilean transformation of velocity would give $u' = 0.6c$.

37.37. Solve: We have

$$p = \frac{mu}{\sqrt{1-u^2/c^2}} = mc \Rightarrow \sqrt{1-u^2/c^2} = \frac{u}{c} \Rightarrow 1 = \frac{2u^2}{c^2} \Rightarrow u = \frac{c}{\sqrt{2}} = 0.707c$$

Assess: The particle's momentum being equal to mc does not mean that the particle is moving with the speed of light. We must use the relativistic formula for the momentum as the particle speeds become high.

37.39. Model: The hamburger is a classical particle whose rest energy is $E_0 = mc^2$.
Solve: **(a)** We have

$$E_0 = mc^2 = (200\times10^{-3}\text{ kg})(3.0\times10^8\text{ m/s})^2 = 1.8\times10^{16}\text{ J}$$

(b) The ratio of the energy equivalent to the food energy is

$$\frac{1.8\times10^{16}\text{ J}}{2\times10^6\text{ J}} = 9.0\times10^9$$

37.43. Model: Let S be the laboratory frame and S′ be the reference frame of the 100 g ball. S′ moves to the right with a speed of $v = 2.0$ m/s relative to frame S. The 50 g ball's speed in frame S is $u_{50} = 4.0$ m/s. Because these speeds are much smaller than the speed of light, we can use the Galilean transformations of velocity.

Visualize:

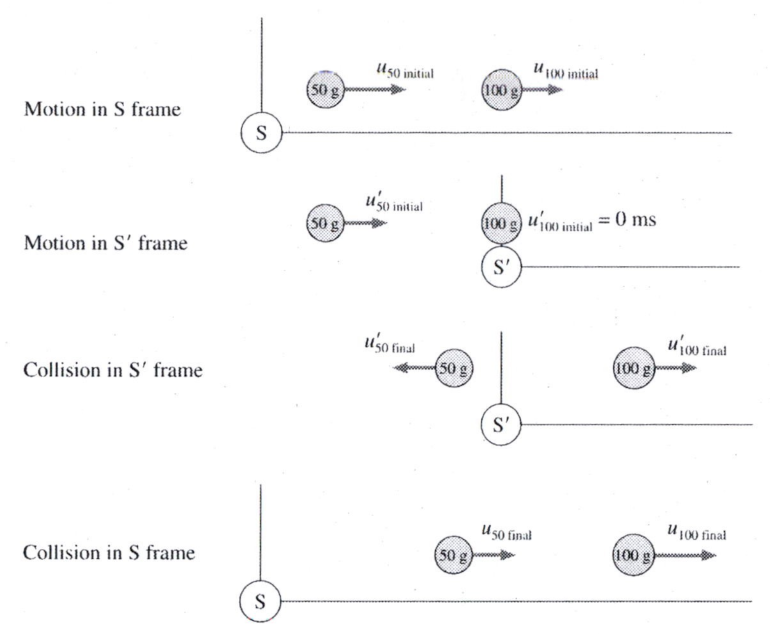

Solve: Transform the collision from frame S into frame S′, where $u'_{100\text{ initial}} = 0$ m/s. Using the Galilean velocity transformation,

$$u'_{50\text{ initial}} = u_{50\text{ initial}} - v = 4.0\text{ m/s} - 2.0\text{ m/s} = 2.0\text{ m/s}$$

Using Equation 10.43,

$$u'_{50\text{ final}} = \frac{50\text{ g} - 100\text{ g}}{50\text{ g} + 100\text{ g}}u'_{50\text{ initial}} = -\left(\frac{1}{3}\right)(2.0\text{ m/s}) = -\left(\frac{2.0}{3}\right)\text{ m/s}$$

$$u'_{100\text{ final}} = \frac{2(50\text{ g})}{50\text{ g} + 100\text{ g}}u'_{50\text{ initial}} = \left(\frac{2}{3}\right)(2.0\text{ m/s}) = +\left(\frac{4.0}{3}\right)\text{ m/s}$$

Using the Galilean transformations of velocity again to go back to the S frame,

$$u_{50\text{ final}} = u'_{50\text{ final}} + v = -\left(\frac{2.0}{3}\right)\text{ m/s} + 2.0\text{ m/s} = +1.33\text{ m/s}$$

$$u_{100\text{ final}} = u'_{100\text{ final}} + v = \left(\frac{4.0}{3}\right)\text{ m/s} + 2.0\text{ m/s} = +3.33\text{ m/s}$$

Because of plus signs with $u_{50\text{ final}}$ and $u_{100\text{ final}}$, both masses are moving to the right.

37.45. Model: Let S be the laboratory frame and S′ be the reference frame of ball 2 after the collision. S′ moves to the right with a speed of $v = 4.0$ m/s relative to the frame S. Because these speeds are much smaller than the speed of light, we can use the Galilean transformation of velocity.

Visualize:

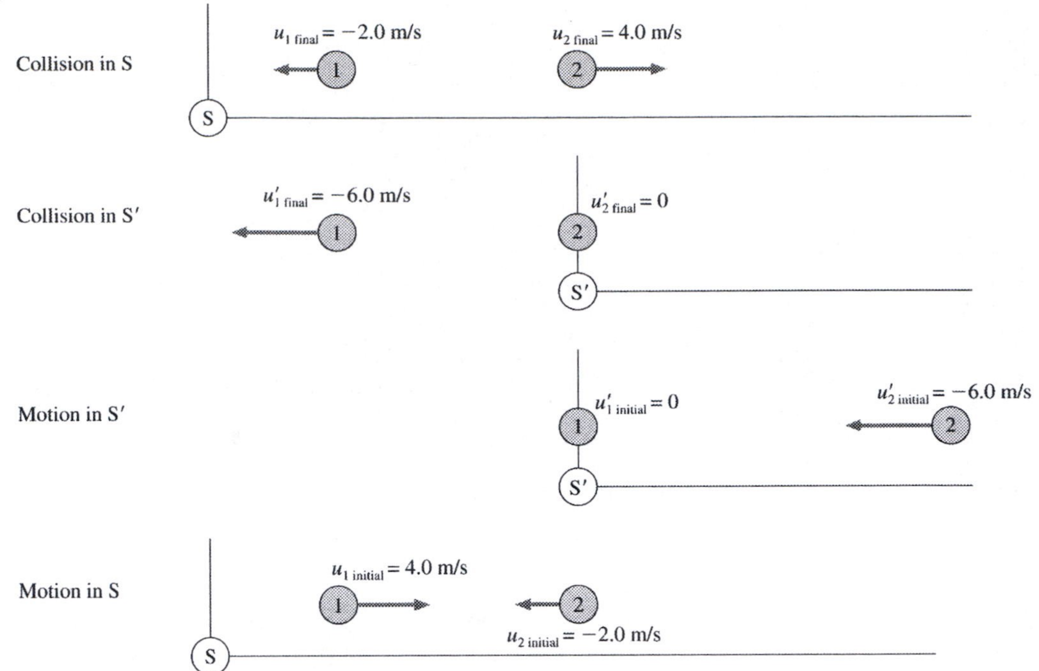

Solve: Transform the collision from frame S into frame S′, where $u'_{2\text{ final}} = 0$ m/s. Using the Galilean velocity transformation,

$$u'_{1\text{ final}} = u_{1\text{ final}} - v = -2.0 \text{ m/s} - 4.0 \text{ m/s} = -6.0 \text{ m/s}$$

In Chapter 10, we found that an elastic collision between two balls of equal mass, where ball 1 was initially at rest ($v_{1i} = 0$), results in $v_{1f} = v_{2i}$ and $v_{2f} = 0$. That is, the balls simply "exchange" velocity. Thus in the S′ frame, the pre-collision velocities must have been $u'_{2\text{ initial}} = u'_{1\text{ final}} = -6.0$ m/s and $u'_{1\text{ initial}} = u'_{2\text{ final}} = 0$ m/s. We can now use the Galilean transformation again to transform the initial velocities in S′ to frame S:

$$u_{1\text{ initial}} = u'_{1\text{ initial}} + v = 0 \text{ m/s} + 4.0 \text{ m/s} = 4.0 \text{ m/s}$$
$$u_{2\text{ initial}} = u'_{2\text{ initial}} + v = -6.0 \text{ m/s} + 4.0 \text{ m/s} = -2.0 \text{ m/s}$$

Before the collision, ball 1 was moving to the right at 4.0 m/s and ball 2 was moving to the left at 2.0 m/s.

37.53. Model: S′ is the electron's frame and S is the ground's frame. S′ moves relative to S with a speed $v = 0.99999997c$.
Solve: For an experimenter in the S frame, the length of the accelerator tube is 3.2 km. This is the proper length $\ell = L$ because it is at rest and is always there for measurements. The electron measures the tube to be length contracted to

$$L' = \sqrt{1-\beta^2}\,\ell = \sqrt{1-(0.99999997)^2}\,(3200 \text{ m}) = 0.78 \text{ m}$$

37.55. Model: Let S be the earth's reference frame and S′ be the rocket's reference frame. S′ travels at $0.5c$ relative to S.
Solve: **(a)** For the earthlings, the total distance traveled by the rocket is 2 × 4.25 ly = 8.5 ly. The time taken by the rocket for the round trip is

$$\frac{8.50 \text{ ly}}{0.5c} = \frac{8.50 \text{ ly}}{0.5 \text{ ly/y}} = 17 \text{ y}$$

(b) The time interval measured in the rocket's frame S′ is the proper time because it can be measured with a single clock at the same position. So,

$$\Delta t = \frac{\Delta\tau}{\sqrt{1-(v/c)^2}} \Rightarrow 17 \text{ y} = \frac{\Delta\tau}{\sqrt{1-(0.5)^2}} \Rightarrow \Delta\tau = 14.7 \text{ y} \approx 15 \text{ y}$$

The distance traveled by the rocket crew is length contracted to

$$L' = L\sqrt{1-(v/c)^2} = (8.50 \text{ ly})\sqrt{1-(0.5)^2} = 7.36 \text{ ly} \approx 7.4 \text{ ly}$$

Note that the speed is still the same:

$$v = \frac{L'}{\Delta\tau} = \frac{7.36 \text{ ly}}{14.7 \text{ y}} = 0.5 \text{ c}$$

(c) Both are correct in their own frame of reference.

37.57. Model: Let S be the earth's reference frame and S′ be the reference frame of one rocket. S′ moves relative to S with $v = -0.75c$. The speed of the second rocket in the frame S is $u = +0.75c$.
Visualize:

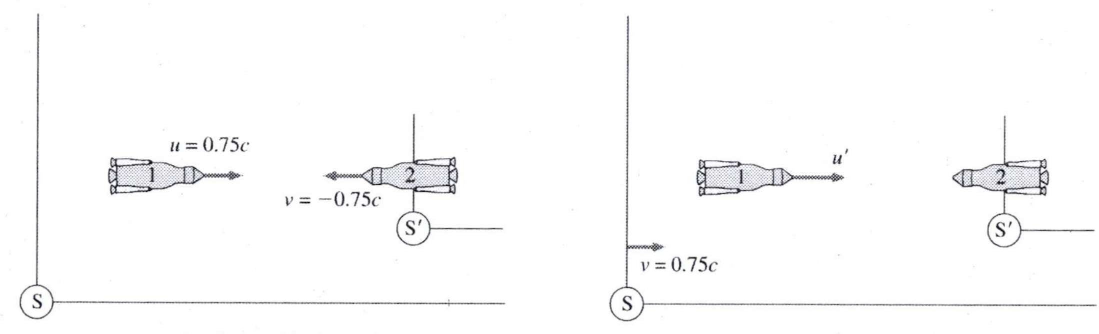

Solve: Using the Lorentz velocity transformation equation,

$$u' = \frac{u - v}{1 - uv/c^2} = \frac{0.75c - (-0.75c)}{1 - (0.75c)(-0.75c)/c^2} = 0.96c$$

Assess: In Newtonian mechanics, the Galilean transformation of velocity will give $u' = 0.75c - (-0.75c) = 1.50c$. This is not permissible according to the theory of relativity.

37.61. Model: Let S be the ground's reference frame and S′ the muon's reference frame. S′ travels with a speed of v relative to S.
Solve: **(a)** The half-life of a muon at rest is 1.5 μs. That is, the half-life in the muon's rest frame S′ is 1.5 μs. So, $\Delta t' = \Delta\tau =$ 1.5 μs. The half-life of 7.5 μs, when muons have been accelerated to very high speed, means that $\Delta t = 7.5\ \mu$s. Thus

$$\Delta t = 7.5\ \mu\text{s} = \frac{\Delta\tau}{\sqrt{1-(v/c)^2}} = \frac{1.5\ \mu\text{s}}{\sqrt{1-v^2/c^2}} \Rightarrow \sqrt{1-v^2/c^2} = 0.20 \Rightarrow v = 0.98c$$

(b) The muon's total energy is

$$E = \gamma_p mc^2 = \frac{1}{\sqrt{1-v^2/c^2}} mc^2 = \left(\frac{1}{0.20}\right)(207)(9.11\times10^{-31} \text{ kg})(3.0\times10^8 \text{ m/s})^2 = 8.5\times10^{-11} \text{ J}$$

37.65. Model: Let S be the earth's frame and S′ the rocket's frame. S′ moves at speed 0.8c relative to S. Also, $u'_y = 0.6c$ and $u'_x = 0$.
Solve: **(a)** Using Equation 36.23 and making note of the relationship $y = y'$,

$$u'_y = \frac{dy'}{dt'} = \frac{dy'}{d\left[\gamma\left(t - vx/c^2\right)\right]} = \frac{dy}{\gamma dt - \frac{v}{c^2}\gamma dx} = \frac{1}{\gamma}\frac{dy/dt}{1 - \frac{v}{c^2}\frac{dx}{dt}} = \frac{1}{\gamma}\frac{u_y}{\left(1 - \frac{u_x v}{c^2}\right)}$$

Similarly

$$u_y = \frac{dy}{dt} = \frac{dy'}{\gamma\left(dt' + \frac{vdx'}{c^2}\right)} = \frac{u'_y}{\gamma\left(1 + \frac{u'_x v}{c^2}\right)}$$

(b) The rocket travels past the earth at $v = 0.8c$. It launches the projectile with velocity components $u_x' = 0$ and $u_y' = 0.6c$. In the earth's frame, the x- and y-components of velocity are

$$u_x = \frac{u_x' + v}{1 + u_x'v/c^2} = \frac{0 + v}{1 + 0} = v = 0.8c$$

$$u_y = \frac{u_y'}{\gamma\left(1 + u_x'v/c^2\right)} = \frac{u_y'}{\gamma(1+0)} = \frac{0.6c}{1/\sqrt{1-(0.80)^2}} = 0.36c$$

Thus the projectile's speed in the earth's frame is

$$u = \sqrt{u_x^2 + u_y^2} = \sqrt{(0.80c)^2 + (0.36c)^2} = 0.877c$$

37.67. Solve: The relationship between energy, momentum, and rest energy is $E^2 - (pc)^2 = (mc^2)^2$. With $E = 4mc^2$, this becomes

$$(4mc^2)^2 - (pc)^2 = (mc^2)^2 \Rightarrow (pc)^2 = 15(mc^2)^2 \Rightarrow p = \sqrt{15}mc = 3.87mc$$

37.71. Model: Mass and energy are equivalent and given by Equation 37.43.
Solve: **(a)** The power plant running at full capacity for 80% of the year runs for

$$(0.80)(365 \times 24 \times 3600)\text{ s} = 2.52 \times 10^7\text{ s}$$

The amount of thermal energy generated per year is

$$3 \times (1000 \times 10^6\text{ J/s}) \times (2.52 \times 10^7\text{ s}) = 7.56 \times 10^{16}\text{ J} \approx 7.6\times10^{16}\text{ J}$$

(b) Since $E_0 = mc^2$, the mass of uranium transformed into thermal energy is

$$m = \frac{E_0}{c^2} = \frac{7.56\times10^{16}\text{ J}}{\left(3.0\times10^8\text{ m/s}\right)^2} = 0.84\text{ kg}$$

37.75. Model: Particles can be created from energy, and particles can return to energy. When a particle and its antiparticle meet, they annihilate each other and create two gamma ray photons.
Solve: The energy of the electron is

$$E_{\text{electron}} = \gamma_p m_e c^2 = \frac{1}{\sqrt{1-(0.9)^2}}\left(9.11\times10^{-31}\text{ kg}\right)\left(3.0\times10^8\text{ m/s}\right)^2 = 1.88 \times 10^{-13}\text{ J}$$

The energy of the positron is the same, so the total energy is $E_{\text{total}} = E_{\text{electron}} + E_{\text{positron}} = 3.76 \times 10^{-13}$ J. The energy is converted to two equal-energy photons. Thus, $E_{\text{total}} = 2hf = 2hc/\lambda$. The wavelength is

$$\lambda = \frac{2hc}{E_{\text{total}}} = \frac{2\left(6.62\times10^{-34}\text{ J s}\right)\left(3\times10^8\text{ m/s}\right)}{3.76\times10^{-13}\text{ J}} = 1.06 \times 10^{-12}\text{ m} \approx 1\text{ pm}$$

Assess: This wavelength is typical of γ-ray photons.

38

THE END OF CLASSICAL PHYSICS

Exercises and Problems

38.3. Model: Assume the fields between the electrodes are uniform and that they are zero outside the electrodes.
Visualize: Without the external magnetic field B, the electrons will be deflected *up* toward the positive electrode. The magnetic field must therefore be directed *out of the page* to exert a balancing downward force on the negative electron.
Solve: In a crossed-field experiment, the magnitudes of the electric and magnetic forces on the electron are given by Equation 38.4. The magnitude of the magnetic field is

$$B = \frac{E}{v} = \frac{\Delta V/d}{v} = \frac{200\text{ V}/8.0\times10^{-3}\text{ m}}{5.0\times10^{6}\text{ m}} = 5.0\times10^{-3}\text{ T}$$

Thus $\vec{B} = (5.0\times10^{-3}\text{ T, out of page})$.

38.5. Model: Assume the electric field ($E = \Delta V/d$) between the plates is uniform.
Visualize: Please refer to Figure 38.9. To balance the weight, the electric force must be directed toward the upper electrode, which is more positive than the lower electrode.
Solve: Since $m_{\text{drop}} = \rho V = \rho\left(\frac{4\pi}{3}\right)R^3$, the equation $m_{\text{drop}}g = q_{\text{drop}}E$ is

$$R^3 = \frac{(\Delta V/d)(15e)}{\frac{4\pi}{3}\rho g} = \frac{(25\text{ V}/0.012\text{ m})(15)(1.60\times10^{-19}\text{ C})}{\frac{4\pi}{3}(860\text{ kg/m}^3)(9.8\text{ m/s}^2)} = 1.4163\times10^{-19}\text{ m}^3 \Rightarrow R = 0.52\ \mu\text{m}$$

38.7. Model: The electron volt is a unit of energy and is defined as the kinetic energy gained by an electron or proton if it accelerates through a potential difference of 1 volt.
Solve: **(a)** Converting electron volts to joules,

$$100\text{ eV} = 100\text{ eV}\times\frac{1.60\times10^{-19}\text{ J}}{1\text{ eV}} = 1.60\times10^{-17}\text{ J}$$

Using the definition of kinetic energy $K = \frac{1}{2}mv^2$,

$$v = \sqrt{\frac{2K}{m}} = \sqrt{\frac{2(1.60\times10^{-17}\text{ J})}{9.11\times10^{-31}\text{ kg}}} = 5.9\times10^{6}\text{ m/s}$$

(b) Likewise, the speed of the neutron is

$$v = \sqrt{\frac{(2)(5.0\text{ MeV})}{1.67\times10^{-27}\text{ kg}}} = \sqrt{\frac{2(8.0\times10^{-13}\text{ J})}{1.67\times10^{-27}\text{ kg}}} = 3.1\times10^{7}\text{ m/s}$$

(c) The mass of the particle is

$$m = \frac{2K}{v^2} = \frac{2(2.09\text{ MeV})}{(1.0\times10^{7}\text{ m/s})^2} = \frac{2(3.34\times10^{13}\text{ J})}{(1.0\times10^{7}\text{ m/s})^2} = 6.69\times10^{-27}\text{ kg}$$

The mass of the particle is the same as the mass of four protons (or two protons and two neutrons). It is an alpha particle.

38.9. Model: The electron volt is a unit of energy. It is defined as the energy gained by an electron if it accelerates through a potential difference of 1 volt.

Solve: **(a)** The kinetic energy is

$$K = \tfrac{1}{2}mv^2 = \tfrac{1}{2}(9.11\times10^{-31}\ \text{kg})(5.0\times10^{6}\ \text{m/s})^2 = 1.139\times10^{-17}\ \text{J}\times\frac{1\ \text{eV}}{1.60\times10^{-19}\ \text{J}} = 71\ \text{eV}$$

(b) The potential energy is

$$U = \frac{1}{4\pi\varepsilon_0}\frac{(e)(-e)}{0.10\ \text{nm}} = \frac{-(9.0\times10^{9}\ \text{N m}^2/\text{C}^2)(1.60\times10^{-19})^2}{0.10\times10^{-9}\ \text{m}} = -2.30\times10^{-18}\ \text{J}\times\frac{1\ \text{eV}}{1.60\times10^{-19}\ \text{J}} = -14\ \text{eV}$$

(c)

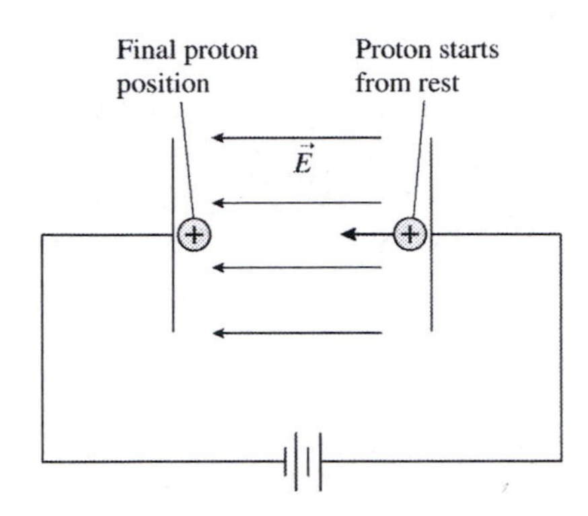

The figure shows a proton accelerating from rest across a parallel-plate capacitor with a potential difference of $\Delta V = 5000$ V. The energy conservation equation $K_f + qV_f = K_i + qV_i$ is

$$K_f = K_i + q(V_i - V_f) = 0\ \text{J} + e\Delta V = e(5000\ \text{V}) = 5000\ \text{eV} = 5.0\ \text{keV}$$

38.17. Model: For a neutral atom, the number of electrons is the same as the number of protons, which is the atomic number Z. An atom's mass number is $A = Z + N$, where N is the number of neutrons.

Solve: **(a)** For a ^{197}Au atom, $Z = 79$. So, $N = 197 - 79 = 118$. A neutral ^{197}Au atom contains 79 electrons, 79 protons, and 118 neutrons.

(b) Assuming that the neutron rest mass is the same as the proton rest mass, the density of the gold nucleus is

$$\rho_{\text{nucleus}} = \frac{197m_{\text{proton}}}{\frac{4\pi}{3}(7.0\times10^{-15}\ \text{m})^3} = \frac{197(1.67\times10^{-27}\ \text{kg})}{\frac{4\pi}{3}(7.0\times10^{-15}\ \text{m})^3} = 2.29\times10^{17}\ \text{kg/m}^3$$

(c) The nuclear density in part (b) is 2.0×10^{13} times the density of lead.

Assess: The mass of the matter is primarily in the nuclei and the volume of the matter is essentially due to the electrons around the nuclei.

38.19. Model: Use Equation 38.11 which is the Balmer formula.

Visualize: Please refer to Figure 38.22.

Solve: **(a)** The wavelengths in the hydrogen emission spectrum are 656.6 nm, 486.3 nm, 434.2 nm, and 410.3 nm. The formula for the Balmer series can be written

$$\frac{1}{m^2} - \frac{1}{n^2} = \frac{91.18\ \text{nm}}{\lambda}$$

where $m = 1, 2, 3, \ldots$ and $n = m + 1, m + 2, \ldots$ For the first wavelength,

$$\frac{1}{m^2} - \frac{1}{n^2} = \frac{91.18\ \text{nm}}{656.5\ \text{nm}} = 0.1389 \Rightarrow \frac{n^2 - m^2}{n^2m^2} = 0.1389$$

This equation is satisfied when $m = 2$ and $n = 3$. For the second wavelength (486.3 nm) the equation is satisfied for $m = 2$ and $n = 4$. Likewise, for the next two wavelengths $m = 2$ and $n = 5$ and 6.

(b) The fifth line in the spectrum will correspond to $m = 2$ and $n = 7$. Its wavelength is

$$\lambda = \frac{91.18\ \text{nm}}{(\frac{1}{2})^2 - (\frac{1}{7})^2} = \left(\frac{196}{45}\right)(91.18\ \text{nm}) = 397.1\ \text{nm}$$

38.27. Model: Use the relativistic expression for the total energy.
Solve: **(a)** The energy of the proton is

$$E = \gamma_p mc^2 = 500\text{ GeV} = \frac{(1.67\times10^{-27}\text{ kg})(3.0\times10^8\text{ m/s})^2}{\sqrt{1-v^2/c^2}} = 500\times10^9\text{ eV}\times\frac{1.60\times10^{-19}\text{ J}}{1\text{ eV}}$$

$$\Rightarrow \sqrt{1-v^2/c^2} = 1.879\times10^{-3} \Rightarrow v = 0.999998c$$

(b) Likewise for the electron,

$$E = 2.0\text{ GeV} = \frac{(9.11\times10^{-31}\text{ kg})(3.0\times10^8\text{m/s})^2}{\sqrt{1-v^2/c^2}}$$

$$\Rightarrow \sqrt{1-v^2/c^2} = 2.562\times10^{-4} \Rightarrow v = 0.99999997c$$

38.31. Model: Assume the fields between the electrodes are uniform and that they are zero outside the electrodes.
Visualize: Please refer to Figures 38.7 and 38.8.
Solve: In a crossed-field experiment, the deflection is zero when the magnetic and electric forces exactly balance each other. From Equation 38.4, the speed of the electrons is

$$v = \frac{E}{B} = \frac{\Delta V/d}{B} = \frac{150\text{ V}/5.0\times10^{-3}\text{ m}}{1.0\times10^{-3}\text{ T}} = 3.0\times10^7\text{ m/s}$$

When the potential difference across the plates is set equal to zero, the magnetic field deflects the electron into a circular path with radius

$$r = \frac{mv}{eB} = \frac{(9.11\times10^{-31}\text{ kg})(3.0\times10^7\text{ m/s})}{(1.6\times10^{-19}\text{ C})(1.0\times10^{-3}\text{ T})} = 0.1708\text{ m}$$

From Figure 38.8, the angle through which the electron is deflected as it passes through the magnetic field is the angle subtended by the arc from where the electron enters the magnetic field to where the electron leaves the magnetic field. Calling this angle θ, we have

$$\sin\theta = \frac{L}{r} = \frac{2.5\text{ cm}}{17.08\text{ cm}} = 0.1463 \Rightarrow \theta = 8.4°$$

Assess: In view of a rather small magnetic field strength, the small deflection angle is reasonable.

38.35. Model: Model the Li atom as a single valence electron orbiting a sphere with net charge $q = +1e$ due to the 3 protons and 2 inner electrons. A sphere of charge acts like a point charge with the total charge concentrated at the center of the sphere.
Visualize:

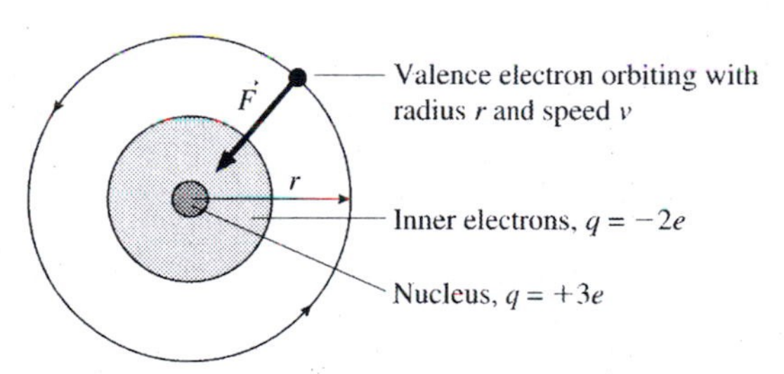

Solve: The electron's energy is both kinetic and potential: $E = \frac{1}{2}mv^2 + kq(-e)/r$. To say that the energy needed to ionize the atom is 5.14 eV means that you would need to increase the electron's energy by 5.14 eV to remove it from the atom, taking it to $r \approx \infty$. Since $E = 0$ for charged particles that are infinitely separated, the energy of the atom must be $E = -5.14\text{ eV} = -8.224\times10^{-19}$ J. Negative energy indicates that the system is bound, and the absolute value is the *binding energy*. Thus from energy considerations we learn that

$$E = \tfrac{1}{2}mv^2 - \frac{Ke^2}{r} = -8.224\times10^{-19}\text{ J}$$

The Coulomb force on the electron provides the centripetal acceleration of circular motion. The force equation is

$$F_r = \frac{Ke^2}{r^2} = ma_r = \frac{mv^2}{r} \Rightarrow \tfrac{1}{2}mv^2 = \frac{Ke^2}{2r}$$

Substitute this expression for the kinetic energy into the energy equation:

$$\tfrac{1}{2}mv^2 - \frac{Ke^2}{r} = \frac{Ke^2}{2r} - \frac{Ke^2}{r} = -\frac{Ke^2}{2r} = -8.224\times10^{-19}\text{ J}$$

$$\Rightarrow r = \frac{(8.99\times10^9\text{ N m}^2/\text{C}^2)(1.60\times10^{-19}\text{ C})^2}{2(8.224\times10^{-19}\text{ J})} = 1.40\times10^{-10}\text{ m} = 0.140\text{ nm}$$

With the radius now known, we can use the result of the force equation to find that

$$v = \sqrt{\frac{Ke^2}{mr}} = 1.34\times10^6\text{ m/s}$$

38.39. Model: The nucleus of an atom contains Z protons and $A - Z$ neutrons.
Solve: **(a)** The values of Z, A, and Z/A for nuclei with $Z = 1, 5, 10, 15, \ldots 90$ are given in the following table.

Z	1	5	10	15	20	25	30	35	40	45	50	55	60	65	70	75	80	85	90
A	1	10.8	20	31	40	55	65	80	91	103	119	133	144	159	173	186	201	210	232
Z/A	1	0.46	0.5	0.48	0.5	0.45	0.46	0.44	0.44	0.44	0.42	0.41	0.42	0.41	0.40	0.40	0.40	0.40	0.39

A graph of Z/A as a function of Z is shown below.

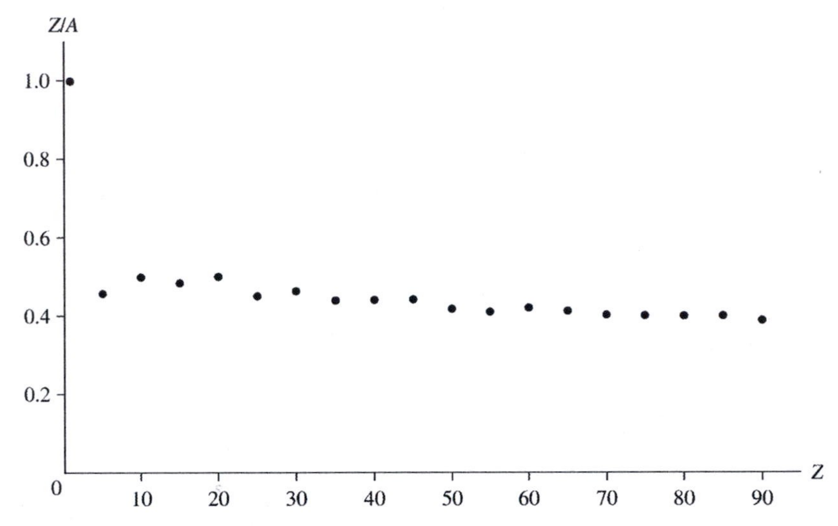

(b) The Z/A values are in the range 0.45–0.50 for values of Z up to approximately 45. For Z in the range 65 to 90, Z/A drops down to approximately 0.40. In short, Z/A decreases from around 0.50 to 0.40 with an increase in Z.
(c) What we are plotting is $Z/Z + N$ as a function of Z. With increasing Z, Z/A decreases because the number of neutrons in the nuclei increases more rapidly than Z.

38.43. Model: Assume the ^{16}O nucleus is at rest. Energy is conserved.
Visualize:

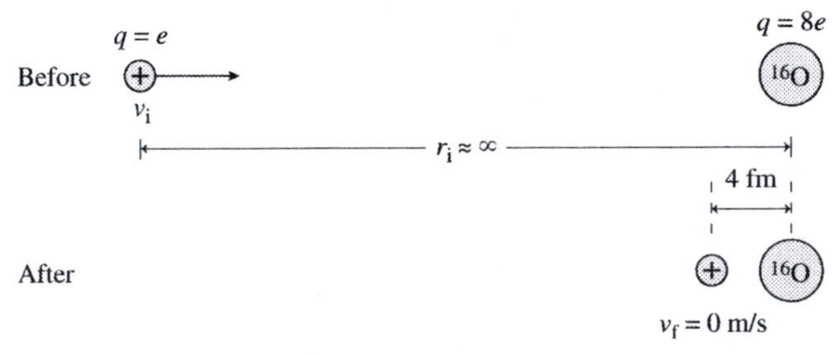

A proton with initial velocity v_i and zero electric potential energy (due to its infinite separation from an ^{16}O nucleus) moves toward ^{16}O and turns around when it is 1.0 fm from the surface. Since r is measured from the center of the nucleus, $r_f = 4$ fm.

Solve: **(a)** The energy conservation equation $K_f + U_f = K_i + U_i$ is

$$0\text{ J} + \frac{1}{4\pi\varepsilon_0}\frac{e(8e)}{(3.0\text{ fm} + 1.0\text{ fm})} = \tfrac{1}{2}mv_i^2 + 0\text{ J}$$

$$\Rightarrow v_i^2 = \frac{(2)(9.0\times10^9\text{ N m}^2/\text{C}^2)(8)(1.6\times10^{-19}\text{ C})^2}{(4.0\times10^{-15}\text{ m})(1.67\times10^{-27}\text{ kg})} = 5.52\times10^{14}\text{ m}^2/\text{s}^2 \Rightarrow v_i = 2.3\times10^7\text{ m/s}$$

(b) $K = \tfrac{1}{2}mv_i^2 = \tfrac{1}{2}(1.67\times10^{-27}\text{ kg})(5.52\times10^{14}\text{ m}^2/\text{s}^2) = 4.60\times10^{-13}\text{ J} = 2.9\text{ MeV}$

39

QUANTIZATION

Exercises and Problems

39.3. Model: Light of frequency f consists of discrete quanta, each of energy $E = hf$.
Solve: The lowest photon energy that creates photoelectrons from the metal is

$$E = \frac{hc}{\lambda} = \frac{(6.63\times10^{-34}\text{ J s})(3.0\times10^{8}\text{ m/s})}{388\times10^{-9}\text{ m}} \times \frac{1\text{ eV}}{1.6\times10^{-19}\text{ J}} = 3.20\text{ eV}$$

The work function of the metal is $E_0 = 3.20$ eV.

39.7. Solve: **(a)** A metal can be identified by its work function. From Equation 39.8, the stopping potential is

$$V_{\text{stop}} = \frac{hf - E_0}{e} \Rightarrow E_0 = hf - eV_{\text{stop}}$$

The frequency and energy of the photons are

$$f = \frac{c}{\lambda} = \frac{3.00\times10^{8}\text{ m/s}}{200\times10^{-9}\text{ m}} = 1.500\times10^{15}\text{ Hz} \Rightarrow hf = (4.14\times10^{-15}\text{ eV s})(1.500\times10^{15}\text{ Hz}) = 6.21\text{ eV}$$

If the stopping potential is $V_{\text{stop}} = 1.93$ V, then $eV_{\text{stop}} = 1.93$ eV. Thus,

$$E_0 = hf - eV_{\text{stop}} = 6.21\text{ eV} - 1.93\text{ eV} = 4.28\text{ eV}$$

Using Table 39.1, we can identify the metal as *aluminum*.
(b) The kinetic energy of the electrons and thus the stopping potential are *independent* of the light intensity. A more intense light generates more electrons, but the electrons still have the same kinetic energy. The stopping potential is $V_{\text{stop}} =$ 1.93 V after the intensity is doubled.

39.11. Solve: **(a)** From Equation 39.4, the energy of each photon is

$$E_{\text{photon}} = hf = (6.63\times10^{-34}\text{ J s})(101\times10^{6}\text{ Hz}) = 6.696\times10^{-28}\text{ J}$$

The number of photons in 10^4 J is

$$N = \frac{E_{\text{total}}}{E_{\text{photon}}} = \frac{10^4\text{ J}}{6.696\times10^{-28}\text{ J}} = 1.5\times10^{29}$$

The antenna emits 1.5×10^{29} photons per second.
(b) The number of photons emitted per second is so enormous that we couldn't possibly recognize the effects of single photons. It's safe to treat the broadcast as an electromagnetic wave.

39.15. Solve: The de Broglie wavelength is $\lambda = h/mv$. Thus,

$$v = \frac{h}{m\lambda} = \frac{6.63\times10^{-34}\text{ J s}}{(9.11\times10^{-31}\text{ kg})(500\times10^{-9}\text{ m})} = 1456\text{ m/s}$$

A potential difference of ΔV will raise the kinetic energy of a rest electron by $\frac{1}{2}mv^2$. Thus,

$$e\Delta V = \frac{1}{2}mv^2 \Rightarrow \Delta V = \frac{mv^2}{2e} = \frac{(9.11\times10^{-31}\text{ kg})(1456\text{ m/s})^2}{2(1.6\times10^{-19}\text{ C})} = 6.0\times10^{-6}\text{ V}$$

Assess: A mere 6.0 × 10^{-6} V is able to increase an electron's speed to 1456 m/s.

39.19. Model: For a "particle in a box," the energy is quantized.
Solve: The energy of the n = 1 state is

$$E_1 = (1)^2\frac{h^2}{8mL^2} = E_{\text{photon}} = \frac{hc}{\lambda} \Rightarrow L = \sqrt{\frac{h\lambda}{8mc}} = \sqrt{\frac{(6.63\times10^{-34}\text{ J s})(600\times10^{-9}\text{ m})}{8(9.11\times10^{-31}\text{ kg})(3.0\times10^{8}\text{ m/s})}} = 0.427\text{ nm}$$

39.21. Model: To conserve energy, the absorption spectrum must have exactly the energy gained by the atom in the quantum jumps.
Visualize: Please refer to Figure EX39.20.
Solve: **(a)** An electron with a kinetic energy of 2.00 eV can collide with an atom in the n = 1 state and raise its energy to the n = 2 state. This is possible because $E_2 - E_1$ = 1.50 eV is less than 2.00 eV. On the other hand, the atom cannot be excited to the n = 3 state.
(b) The atom will absorb 1.50 eV of energy from the incoming electron, leaving the electron with 0.50 eV of kinetic energy.

39.23. Model: The electron must have $k \geq \Delta E_{\text{atom}}$ to cause collisional excitation. The atom is initially in the n = 1 ground state.
Visualize:

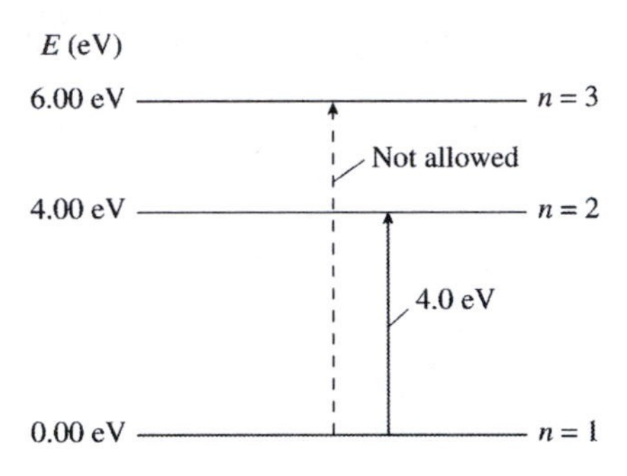

Solve: The kinetic energy of the incoming electron is

$$E = \tfrac{1}{2}mv^2 = \tfrac{1}{2}(9.11\times10^{-31}\text{ kg})(1.30\times10^{6}\text{ m/s})^2 = 7.698\times10^{-19}\text{ J} = 4.81\text{ eV}$$

The electron has enough energy to excite the atom to the n = 2 stationary state ($E_2 - E_1$ = 4.00 eV). However, it does not have enough energy to excite the atom into the n = 3 state which requires a total energy of 6.00 eV.

39.25. Solve: **(a)** From Equation 39.25, the radius in state n of a hydrogen atom is $r_n = n^2a_B$. A 100 nm diameter atom has r_n = 50 nm. Thus, the quantum state is

$$n = \sqrt{\frac{r_n}{a_B}} = \sqrt{\frac{50\text{ nm}}{0.0529\text{ nm}}} = 30.74$$

Since n has to be an integer, we obtain n = 31.
(b) From Equation 39.27, the electron's speed is

$$v_{31} = \frac{v_1}{31} = \frac{2.19\times10^{6}\text{ m/s}}{31} = 7.06\times10^{4}\text{ m/s}$$

From Equation 39.30, the electron's energy is

$$E_{31} = -\frac{E_1}{(31)^2} = -\frac{13.60\text{ eV}}{(31)^2} = -0.0142\text{ eV}$$

39.33. Solve: For hydrogen-like ions, Equation 39.37 is

$$r_n = n^2 \frac{a_B}{Z} = \frac{(r_n)_H}{Z} \qquad v_n = Z\frac{v_1}{n} = Z \cdot (v_n)_H \qquad E_n = -Z^2\left(\frac{13.60 \text{ eV}}{n^2}\right) = Z^2(E_n)_H$$

Where $(r_n)_H$, $(v_n)_H$, and $(E_n)_H$ are the values of ordinary hydrogen. He^+ has $Z = 2$. Using Table 39.2 for the values of hydrogen, we get

n	r_n (nm)	v_n (m/s)	E_n (eV)
1	0.026	4.38×10^6	−54.4
2	0.106	2.19×10^6	−13.6
3	0.238	1.46×10^6	−6.0

39.37. Solve: **(a)** The threshold frequency is $f_0 = E_0/h$. The threshold frequency for potassium and gold are given in the table in part (b).
(b) The corresponding threshold wavelength is $\lambda_0 = c/f_0$. The results of the calculations are in the table below.

Metal	E_0 (eV)	f_0 (Hz)	λ_0 (nm)
Potassium	2.30	5.56×10^{14}	540
Gold	5.10	1.23×10^{15}	244

(c) When light of wavelength λ is incident on the metal, the maximum kinetic energy of the photoelectrons is

$$K_{max} = \tfrac{1}{2}mv_{max}^2 = hf - E_0 = \frac{hc}{\lambda} - E_0 \Rightarrow v_{max} = \sqrt{\frac{2}{m}\left(\frac{hc}{\lambda} - E_0\right)}$$

E_0 must be converted to SI units of joules before this formula can be used. v_{max} for potassium and gold are given in the table in part (d).
(d) The stopping potential is

$$V_{stop} = \frac{K_{max}}{e} = \frac{1}{e}\left(\frac{hc}{\lambda} - E_0\right)$$

where again E_0 has to be joules. The results of the calculations are in the table below.

Metal	E_0 (J)	v_{max} (m/s)	V_{stop} (V)
Potassium	3.68×10^{-19}	10.8×10^5	3.35
Gold	8.16×10^{-19}	4.4×10^5	0.55

39.39. Solve: The stopping potential is $eV_{stop} = hf - E_0$. Substituting the given values,

$$eV_{stop} = \frac{hc}{300 \text{ nm}} - E_0 \qquad 0.257(eV_{stop}) = \frac{hc}{400 \text{ nm}} - E_0$$

Substituting the first equation into the second,

$$0.257\left(\frac{hc}{300 \text{ nm}} - E_0\right) = \frac{hc}{400 \text{ nm}} - E_0 \Rightarrow E_0(1 - 0.257) = hc\left(\frac{1}{400 \text{ nm}} - \frac{0.257}{300 \text{ nm}}\right)$$

$$\Rightarrow E_0 = \frac{(4.14\times10^{-15} \text{ eV s})(3.0\times10^8 \text{ m/s})(1.643\times10^{-3}/\text{nm})}{0.743} = 2.75 \text{ eV}$$

From Table 39.1, we identify the cathode metal as sodium.

39.41. Solve: **(a)** The cathode is illuminated with light of wavelength 300 nm. The frequency of the light is

$$f = \frac{c}{\lambda} = \frac{3.0\times10^8 \text{ m/s}}{300\times10^{-9} \text{ m}} = 1.0\times10^{15} \text{ Hz}$$

To plot a graph of electron current I as a function of the potential difference ΔV, we need to know (i) the stopping potential and (ii) the saturation current. The equation for the stopping potential is

$$V_{stop} = \frac{h}{e}f - \frac{h}{e}f_0$$

According to the data shown in Figure P39.41, the slope of the graph is

$$\text{slope} = \frac{h}{e} = \frac{2.5\ \text{V}}{6.0\times10^{14}\ \text{Hz}} = 4.167\times10^{-15}\ \text{V s}$$

Also, f_0 is the threshold frequency and its value is 6.0×10^{14} Hz. The stopping potential at $f = 1.0 \times 10^{15}$ Hz is

$$V_{\text{stop}} = (4.167 \times 10^{-15}\ \text{Vs})(1.0 \times 10^{15}\ \text{Hz}) - (4.167 \times 10^{-15}\ \text{V s})(6.0 \times 10^{14}\ \text{Hz}) = 1.67\ \text{V}$$

The rate of photons arriving at the cathode is

$$R = \frac{P}{hf} = \frac{10\times10^{-6}\ \text{W}}{(6.63\times10^{-34}\ \text{J s})(1.0\times10^{15}\ \text{Hz})} = 1.508\times10^{13}\ \text{photons/s}$$

If 10% of the photons eject an electron, the electron current is $i = 1.508 \times 10^{12}$ electrons/s. Thus the actual current between the cathode and anode is

$$I = ei = (1.60 \times 10^{-19}\ \text{C/electron})(1.508 \times 10^{12}\ \text{electrons/s}) = 2.4 \times 10^{-7}\ \text{A} = 0.24\ \mu\text{A}$$

For positive anode voltages, the current will level off at this value when all photoelectrons are collected at the anode. For negative voltages, the current will decrease until stopping at –1.67 V. Thus the current-versus-voltage graph looks as follows.

(b)

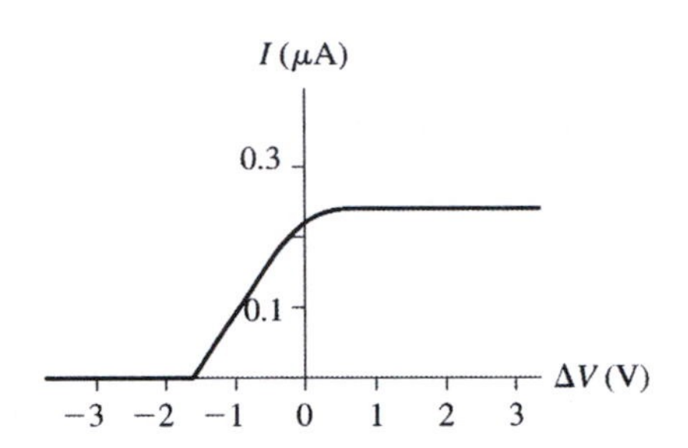

39.45. Model: Neutrons have both particle-like and wave-like properties.

Visualize:

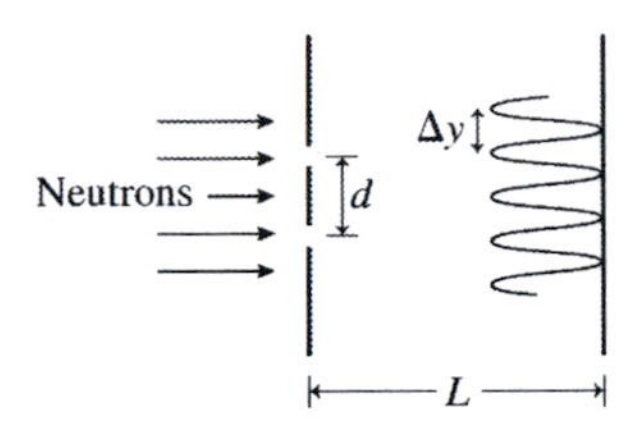

Solve: **(a)** The kinetic energy of the neutron is

$$K = \tfrac{1}{2}mv^2 = \tfrac{1}{2}\left(1.67\times10^{-27}\ \text{kg}\right)\left(200\ \text{m/s}\right)^2 = 3.34\times10^{-23}\ \text{J} = 2.1\times10^{-4}\ \text{eV}$$

(b) The de Broglie wavelength at this speed is

$$\lambda = \frac{h}{mv} = \frac{6.63\times10^{-34}\ \text{J s}}{\left(1.67\times10^{-27}\ \text{kg}\right)\left(200\ \text{m/s}\right)} = 2.0\times10^{-9}\ \text{m} = 2.0\ \text{nm}$$

(c) From Equation 22.7, the fringe spacing in a double-slit interference experiment is $\Delta y = \lambda L/d$, where d is the slit separation and L is the distance to the detector. From Figure P39.45, the spacing between the two peaks with $m = \pm1$ (on either side of the central maximum) is 1.4 times as long as the length of the reference bar, which gives $\Delta y = 70\ \mu\text{m}$.

Thus, the distance from the slits to the detector was

$$L = \frac{d\Delta y}{\lambda} = \frac{\left(1.0\times10^{-4}\ \text{m}\right)\left(7.0\times10^{-5}\ \text{m}\right)}{2.0\times10^{-9}\ \text{m}} = 3.5\ \text{m}$$

39.51. Model: Photons are emitted when an atom undergoes a quantum jump from a higher energy level to a lower energy level. On the other hand, photons are absorbed in a quantum jump from a lower energy level to a higher energy level. Because most of the atoms are in the $n = 1$ ground state, the only quantum jumps in the absorption spectrum start from the $n = 1$ state.

Solve: **(a)** The ionization energy is $|E_1| = 6.5$ eV.

(b) The absorption spectrum consists of the transitions $1 \to 2$ and $1 \to 3$ from the ground state to excited states. According to the Bohr model, the required photon frequency and wavelength are

$$f = \frac{\Delta E}{h} \Rightarrow \lambda = \frac{c}{f} = \frac{hc}{\Delta E}$$

where $\Delta E = E_f - E_i$ is the energy change of the atom. Using the energies given in the figure, we calculated the values in the table below.

Transition	E_f (eV)	E_i (eV)	ΔE (eV)	λ (nm)
$1 \to 2$	−3.0	−6.5	3.5	355
$1 \to 3$	−2.0	−6.5	4.5	276

(c) Both wavelengths are ultraviolet ($\lambda < 400$ nm).

(d) A photon with wavelength $\lambda = 1240$ nm has an energy $E_{photon} = hf = hc/\lambda = 1.0$ eV. Because E_{photon} must exactly match ΔE of the atom, a 1240 nm photon can be emitted only in a $3 \to 2$ transition. So, after the collision the atom was in the $n = 3$ state. Before the collision, the atom was in its ground state ($n = 1$). Thus, an electron with $v_i = 1.4 \times 10^6$ m/s collided with the atom in the $n = 1$ state. The atom gained 4.5 eV in the collision as it is was excited from the $n = 1$ to $n = 3$, so the electron lost 4.5 eV = 7.20×10^{-19} J of kinetic energy. Initially, the kinetic energy of the electron was

$$K_i = \tfrac{1}{2} m_{elec} v_i^2 = \tfrac{1}{2}(9.11 \times 10^{-31} \text{ kg})(1.40 \times 10^6 \text{ m/s})^2 = 8.93 \times 10^{-19} \text{ J}$$

After losing 7.20×10^{-19} J in the collision, the kinetic energy is

$$K_f = K_i - 7.20 \times 10^{-19} \text{ J} = 1.73 \times 10^{-19} \text{ J} = \tfrac{1}{2} m_{elec} v_f^2 \Rightarrow v_f = \sqrt{\frac{2K_f}{m_{elec}}} = \sqrt{\frac{2(1.73 \times 10^{-19} \text{ J})}{9.11 \times 10^{-31} \text{ kg}}} = 6.16 \times 10^5 \text{ m/s}$$

39.57. Solve: **(a)** The orbital radius and velocity for $n = 99$ are,

$$r_{99} = 99^2 a_B = 518 \text{ nm} \qquad v_{99} = \frac{v_1}{99} = 2.21 \times 10^4 \text{ m/s}$$

For $n = 100$,

$$r_{100} = 100^2 a_B = 529 \text{ nm} \qquad v_{100} = \frac{v_1}{100} = 2.19 \times 10^4 \text{ m/s}$$

(b) For circular motion, the orbital frequency is $f = v/2\pi r$. Using r and v from part (a),

$$f_{99} = 6.79 \times 10^9 \text{ Hz} \qquad f_{100} = 6.59 \times 10^9 \text{ Hz}$$

(c) Using Equation 38.36, the wavelength of a photon emitted in a $100 \to 99$ transition is

$$\lambda_{100 \to 99} = \frac{91.18 \text{ nm}}{(99)^{-2} - (100)^{-2}} = 4.49 \times 10^7 \text{ nm} = 0.0449 \text{ m}$$

The frequency of this photon is $f_{photon} = c/\lambda = 6.68 \times 10^9$ Hz.

(d) The average of f_{99} and f_{100} from part (b) is $f_{avg} = 6.69 \times 10^9$ Hz. This differs from the photon frequency f_{photon} by only 0.01×10^9 Hz, or 0.15%.

39.59. Visualize: A neutral oxygen atom that has lost 7 electrons is a hydrogen-like atom because it has one electron going around a nucleus with $Z = 8$.
Solve: The energy levels for the O^{+7} ion are

$$E_n = -\frac{13.60\,Z^2\text{ eV}}{n^2} = -\frac{13.60(8)^2\text{ eV}}{n^2} = -\frac{870.4\text{ eV}}{n^2}$$

The energy of the $3 \rightarrow 2$ transition is

$$E_3 - E_2 = -870.4\text{ eV}\left(\frac{1}{3^2} - \frac{1}{2^2}\right) = 120.89\text{ eV} = \frac{hc}{\lambda}$$

$$\Rightarrow \lambda = \frac{(4.14\times10^{-15}\text{ eV})(3.0\times10^{8}\text{ m/s})}{120.89\text{ eV}} = 10.28\text{ nm}$$

Likewise for the $4 \rightarrow 2$ transition,

$$E_4 - E_2 = -870.4\text{ eV}\left(\frac{1}{4^2} - \frac{1}{2^2}\right) = 163.2\text{ eV}$$

The wavelength for this transition $\lambda = 7.62$ nm. For the $5 \rightarrow 2$ transition, $E_5 - E_2 = 182.78$ eV and $\lambda = 6.80$ nm. All the wavelengths are in the ultraviolet range.

40

WAVE FUNCTIONS AND UNCERTAINTY

Exercises and Problems

40.5. Model: The probability that the outcome will be A or B is the sum of P_A and P_B.

Solve: (a) Each die has six faces and the faces have dots numbering from 1 to 6. We have two dice A and B. The various possible outcomes of rolling two dice are given in the following table.

A	B	A	B	A	B
1	1	3	1	5	1
1	2	3	2	5	2
1	3	3	3	5	3
1	4	3	4	5	4
1	5	3	5	5	5
1	6	3	6	5	6
2	1	4	1	6	1
2	2	4	2	6	2
2	3	4	3	6	3
2	4	4	4	6	4
2	5	4	5	6	5
2	6	4	6	6	6

There are 36 possible outcomes. From the table, we find that there are six ways of rolling a 7 (1 and 6, 2 and 5, 3 and 4, 4 and 3, 5 and 2, 6 and 1). The probability is $(1/36) \times 6 = 1/6$.

(b) Likewise, the probability of rolling a double is 1/6.

(c) There are 10 ways of rolling a 6 or an 8. The probability is $(1/36) \times 10 = 5/18$.

40.7. Visualize: Combine Equations 40.10 and 40.11 to show that N is proportional to $|A(x)|^2\,\delta x$.

$$\frac{|A(x_2)|^2\,\delta x_2}{|A(x_1)|^2\,\delta x_1} = \frac{\dfrac{N(\text{in } \delta x_2 \text{ at } x_2)}{N_{\text{tot}}}}{\dfrac{N(\text{in } \delta x_1 \text{ at } x_1)}{N_{\text{tot}}}}$$

We are given $N_1 = 6000$, $\delta x_1 = 0.10$ mm, $A(x_1) = 200$ V/m, $N_2 = 3000$, and $\delta x_2 = 0.20$ mm. We are not given N_{tot} but it cancels anyway.

Solve: Solve the above equation for $|A(x_2)|$.

$$|A(x_2)| = |A(x_1)|\sqrt{\frac{\delta x_1 N(\text{in } \delta x_2 \text{ at } x_2)}{\delta x_2 N(\text{in } \delta x_1 \text{ at } x_1)}} = (200\text{ V/m})\sqrt{\frac{(0.10\text{ mm})(3000)}{(0.20\text{ mm})(6000)}} = 100\text{ V/m}$$

Assess: The answer is half of the wave amplitude at the other strip, which seems reasonable.

40.11. Model: The probability of finding a particle at position x is determined by $|\psi(x)|^2$.

Solve: **(a)** The probability of detecting an electron is Prob(in δx at x) = $|\psi(x)|^2\,\delta x$. At x = 0 mm, the number of electrons landing is calculated as follows:

$$\frac{N}{1.0\times10^6}=|\psi(0\text{ mm})|^2\,\delta x \Rightarrow N=\left(\tfrac{1}{3}\text{ mm}^{-1}\right)(0.010\text{ mm})\left(1.0\times10^6\right)=3333$$

(b) Likewise, the number of electrons landing at x = 2.0 mm is

$$N=|\psi(2.0\text{ mm})|^2\,\delta x N_{\text{total}}=\left(0.111\text{ mm}^{-1}\right)(0.010\text{ mm})\left(1.0\times10^6\right)=1111$$

40.13. Model: The probability of finding a particle at position x is determined by $|\psi(x)|^2$.

Solve: **(a)** The probability of detecting an electron is $\text{Prob}(\text{in }\delta x\text{ at }x)=|\psi(x)|^2\,\delta x$. Hence the probability the electron will land at x = 0.000 mm is

$$\text{Prob(in 0.010 mm at } x = 0.000\text{ mm}) = (0.50\text{ mm}^{-1})(0.010\text{ mm}) = 5.0\times10^{-3}$$

(b) Since $|\psi(0.500\text{ mm})|^2 = 0.25$ mm, the probability is 2.5×10^{-3}.

(c) Since $|\psi(1.000\text{ mm})|^2 = 0$, the probability is 0.

(d) Since $|\psi(2.000\text{ mm})|^2 = 0.25\text{ mm}^{-1}$, the probability is 2.5×10^{-3}.

40.15. Model: The probability of finding the particle is determined by the probability density $P(x)=|\psi(x)|^2$.

Solve: **(a)** According to Equation 40.18, $\int_{-\infty}^{\infty}|\psi(x)|^2\,dx$ = area under the curve = 1. The area of the $|\psi(x)|^2$-versus-x graph is $4a$ mm. Hence, $a=\tfrac{1}{4}\text{ mm}^{-1}$.

(b) Each point on the $\psi(x)$ graph is the square root of the corresponding point on the $|\psi(x)|^2$ graph. Where the $|\psi(x)|^2$ graph has reached 1/2 its maximum value at x = 2 mm, the $\psi(x)$ graph will have reached to $1/\sqrt{2}$ = 0.707 of its maximum value. Thus the graph shape is convex. Since $a=\tfrac{1}{4}\text{ mm}^{-1}$, the maximum value of $\psi(x)$ is $\sqrt{a}=1/\sqrt{4}\text{ nm}^{-1/2}=0.50\text{ mm}^{-1/2}$. The graph is shown below. The negative of this graph, curving from negative to positive, would also be an acceptable wave function.

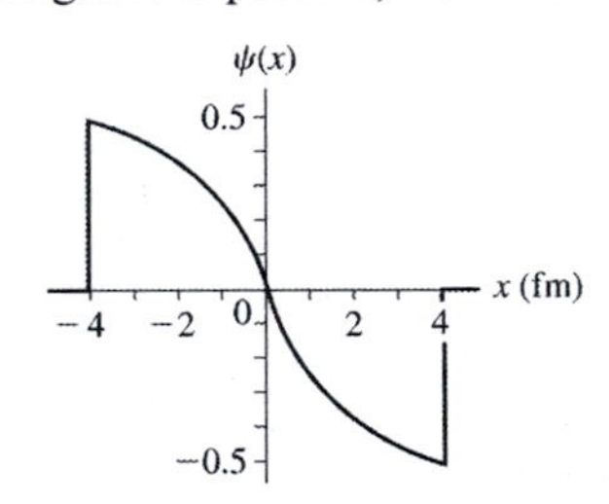

(c) The probability of the neutron being located in the interval $|x|\geq 2.0$ fm is

$$\text{Prob}(|x|\geq 2.0\text{ fm}) = 1 - \text{area under the curve between } -2.0\text{ fm and } 2.0\text{ fm}\omega_{0z}$$

$$=1-2\times\frac{1}{2}\left(\frac{a}{2}\right)(2.0\text{ fm}-0\text{ fm})=1-2\times\frac{(0.25\text{ fm}^{-1})(2.0\text{ fm})}{4}=0.75$$

40.21. Model: A laser pulse is an electromagnetic wave packet, hence it must satisfy the relationship $\Delta f\,\Delta t\approx 1$.

Solve: Because $c=\lambda f$, the frequency and period are

$$f=\frac{3.0\times10^8\text{ m/s}}{1.5\times10^{-6}\text{ m}}=2.0\times10^{14}\text{ Hz}\Rightarrow T=\frac{1}{f}=\frac{1}{2.0\times10^{14}\text{ Hz}}=5.0\times10^{-15}\text{ s}$$

Since Δf = 2.0 GHz, the minimum pulse duration is

$$\Delta t\approx\frac{1}{\Delta f}=\frac{1}{2.0\times10^9\text{ Hz}}=5.0\times10^{-10}\text{ s}$$

The number of oscillations in this laser pulse is

$$\frac{5.0\times10^{-10}\ \text{s}}{T}=\frac{5.0\times10^{-10}\ \text{s}}{5.0\times10^{-15}\ \text{s}}=1.0\times10^{5}\ \text{oscillations}$$

40.25. Model: Protons are subject to the Heisenberg uncertainty principle.
Solve: We know the proton is somewhere within the nucleus, so the uncertainty in our knowledge of its position is at most $\Delta x = L = 4.0$ fm. With a finite Δx, the uncertainty Δp_x is given by the uncertainty principle:

$$\Delta p_x = m\Delta v_x = \frac{h/2}{\Delta x} \Rightarrow \Delta v_x = \frac{h}{2mL} = \frac{6.63\times10^{-34}\ \text{J s}}{2\left(1.67\times10^{-27}\ \text{kg}\right)\left(4.0\times10^{-15}\ \text{m}\right)} = 5.0\times10^{7}\ \text{m/s}$$

Because the average velocity is zero, the best we can say is that the proton's velocity is somewhere in the range -2.5×10^{7} m/s to 2.5×10^{7} m/s. Thus the smallest range of speeds is 0 to 2.5×10^{7} m/s.

40.31. Model: The probability of finding a particle at position x is determined by $P(x)=|\psi(x)|^2$.
Visualize:

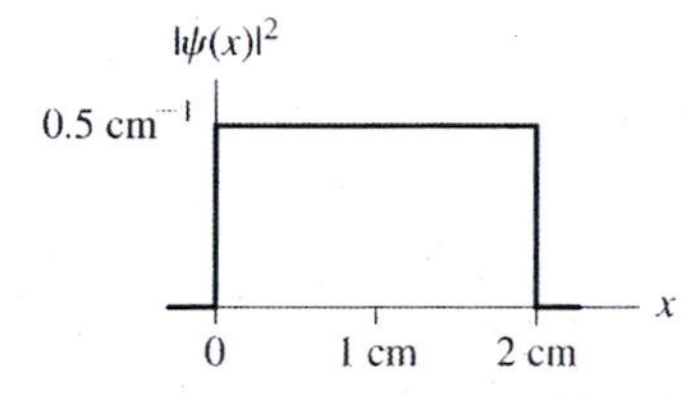

Solve: **(a)** Since the electrons are uniformly distributed over the interval $0 \le x \le 2$ cm, the probability density $P(x)=|\psi(x)|^2$ is constant over this interval. $P(x) = 0$ outside this interval because no electrons are detected. Thus $|\psi(x)|^2$ is a square function, as shown in the figure. To be normalized, the area under the probability curves must be 1. Hence, the peak value of $|\psi(x)|^2$ must be 0.5 cm^{-1}.
(b) The interval is $\delta x = 0.02$ cm. The probability is

$$\text{Prob(in } \delta x \text{ at } x = 0.80\ \text{cm)} = |\psi(x=0.80\ \text{cm})|^2\,\delta x = (0.5\ \text{cm}^{-1})(0.02\ \text{cm}) = 0.01 = 1\%$$

(c) From Equation 39.7, the number of electrons is

$$N(\text{in } \delta x \text{ at } x = 0.80\ \text{cm}) = N_{\text{total}}\text{Prob(in } \delta x \text{ at } x = 0.80\ \text{cm)} = 10^{6} \times (0.01) = 10^{4}$$

(d) The probability density is $P(x = 0.80\ \text{cm}) = |\psi(x = 0.80\ \text{cm})|^2 = 0.5\ \text{cm}^{-1}$.

40.33. Model: The probability of finding a particle at position x is determined by $P(x)=|\psi(x)|^2$.
Visualize:

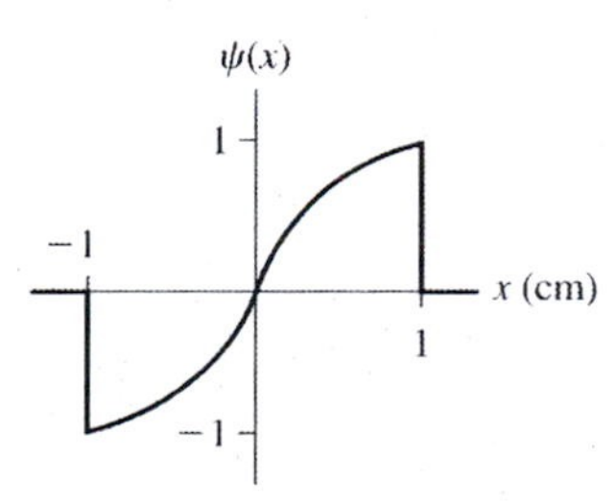

Solve: **(a)** Yes, because the area under the $|\psi(x)|^2$ curve is equal to 1.
(b) There are two things to consider when drawing $\psi(x)$. First $\psi(x)$ is an *oscillatory* function that changes sign every time it reaches zero. Second, $\psi(x)$ must have the right shape. Each point on the $\psi(x)$ curve is the square root of the corresponding point on the $|\psi(x)|^2$ curve. The values $|\psi(x)|^2 = 1\ \text{cm}^{-1}$ and $|\psi(x)|^2 = 0\ \text{cm}^{-1}$ clearly give $\psi(x) = \pm1\ \text{cm}^{-1/2}$ and $\psi(x) = 0\ \text{cm}^{-1/2}$, respectively. But consider $x = 0.5$ cm, where $|\psi(x)|^2 = 0.5\ \text{cm}^{-1}$. Because $\sqrt{0.5} = 0.707$, $\psi(x = 0.5\ \text{cm}) = 0.707\ \text{cm}^{-1/2}$. This tells us that the $\psi(x)$ curve is not linear but *bows upward* over the interval $0 \le x \le 1$ cm. Thus, $\psi(x)$ has the shape shown in the above figure.

(c) δx = 0.0010 cm is a very small interval, so we can use Prob(in δx at x) = $|\psi(x)|^2\,\delta x$. The values of $|\psi(x)|^2$ can be read from Figure P40.33. Thus,

$$\text{Prob(in } \delta x \text{ at } x = 0.0 \text{ cm)} = |\psi(x = 0.0 \text{ cm})|^2\,\delta x = (0.0 \text{ cm}^{-1})(0.0010 \text{ cm}) = 0.000$$

$$\text{Prob(in } \delta x \text{ at } x = 0.5 \text{ cm)} = |\psi(x = 0.5 \text{ cm})|^2\,\delta x = (0.5 \text{ cm}^{-1})(0.0010 \text{ cm}) = 0.0005$$

$$\text{Prob(in } \delta x \text{ at } x = 0.999 \text{ cm)} = |\psi(x = 1.0 \text{ cm})|^2\,\delta x = (1.0 \text{ cm}^{-1})(0.0010 \text{ cm}) = 0.0010$$

(d) The number of electrons in the interval $-0.3 \text{ cm} \le x \le 0.3$ cm is

$$N(\text{in } -0.3 \text{ cm} \le x\ 0.3 \text{ cm}) = N_{\text{total}} \times \text{Prob(in } -0.3 \text{ cm} \le x \le 0.3 \text{ cm)}$$

The probability is the area under the probability density curve. We have

$$\text{Prob(in } -0.3 \text{ cm} \le x \le 0.3 \text{ cm)} = \int_{-0.3 \text{ cm}}^{0.3 \text{ cm}} |\psi(x)|^2 dx = 2\times\left(\tfrac{1}{2}\times 0.3 \text{ cm}\times 0.3 \text{ cm}^{-1}\right) = 0.090$$

Thus, the number of electrons expected to land in the interval $-0.3 \text{ cm} \le x \le 0.3$ cm is 10,000 × 0.090 = 900.

40.37. Model: The probability of finding a particle at position x is determined by the probability density $P(x) = |\psi(x)|^2$.

Solve: **(a)** The wave function $\psi(x)$ = $(1.414 \text{ nm}^{-1/2})e^{-x/1 \text{ nm}}$ changes over a length scale of ≈ 1 nm. The distance δx = 0.01 nm is very small compared to 1 nm. So we can use

$$\text{Prob(in } \delta x = 0.01 \text{ nm at } x = 1 \text{ nm)} = |\psi(x = 1 \text{ nm})|^2\,\delta x$$

$$= \left[(1.414 \text{ nm}^{-1/2})e^{-1}\right]^2 (0.01 \text{ nm}) = 2e^{-2}(0.01) = 0.0027 = 0.27\%$$

(b) The interval $0.5 \text{ nm} \le x \le 1.5$ nm is *not* small compared to 1 nm, so we'll need to integrate

$$\text{Prob}(0.5 \text{ nm} \le x \le 1.5 \text{ nm}) = \int_{0.5}^{1.5} |\psi(x)|^2 dx = 2\int_{0.5}^{1.5} e^{-2x}dx = -e^{-2x}\Big|_{0.5}^{1.5} = 0.318 = 31.8\%$$

40.41. Model: The probability of finding a particle at position x is determined by the probability density $P(x) = |\psi(x)|^2$.

Solve: **(a)** The given probability density means that $\psi_1(x) = \sqrt{a/(1-x)}$ in the range -1 mm $\le x \le 0$ mm and $\psi_2(x) = \sqrt{b(1-x)}$ in the range 0 mm $\le x \le 1$ mm. Because $\psi_1(x = 0 \text{ mm}) = \psi_2(x = 0 \text{ mm})$, $\sqrt{a} = \sqrt{b}$ and thus $a = b$.

(b) At $x = -1$ mm, $P(x) = \frac{1}{2}a$; at x = 0 mm, $P(x) = a$; and at x = 1 mm, $P(x)$ = 0. Furthermore, $P(x)$ is a linear function of x for $0 \le x \le 1$ mm.

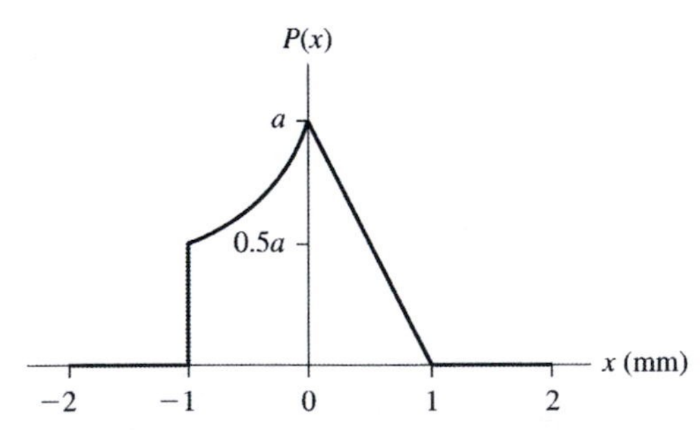

(c) Normalization of the wave function requires integrating over the entire range. We have

$$\int_{-\infty}^{\infty} P(x)dx = \int_{-1}^{0}\left(\frac{a}{1-x}\right)dx + \int_{0}^{1} b(1-x)dx = -a\left[\ln(1+x)\right]_{-1}^{0} + b\left[x - \frac{x^2}{2}\right]_0^1 = a\ln 2 + \frac{b}{2} = a\ln 2 + \frac{a}{2} = a\left(\ln 2 + \tfrac{1}{2}\right)$$

where we have used $a = b$ from part (a). Because $\int P(x)dx = 1$, we have $a = b = 1/(\ln 2 + \frac{1}{2}) = 1/1.193 = 0.838$.

(d) The probability is

$$\int_{-1}^{0} P(x)dx = \int_{-1}^{0} \frac{a}{1-x}dx = -a\left[\ln(1-x)\right]_{-1}^{0} = a\ln 2 = \frac{1}{1.193}(0.693) = 0.581 = 58.1\%$$

40.45. Model: An atom is a particle and is thus subject to the Heisenberg uncertainty principle.
Solve: (a) Since the atom is confined within a box 1 mm in length, the uncertainty in our knowledge of its position is Δx = 1 mm. The uncertainty in the atom's momentum and velocity are

$$\Delta p_x = \frac{h/2}{\Delta x} = \frac{6.63\times10^{-34}\text{ J s}}{2\left(1\times10^{-3}\text{ m}\right)} = 3.32\times10^{-31}\text{ kg m/s}$$

$$\Delta v_x = \frac{\Delta p_x}{m} = \frac{3.315\times10^{-31}\text{ kg m/s}}{23\times1.67\times10^{-27}\text{ kg}} = 8.6\times10^{-6}\text{ m/s}$$

This range of possible velocities will be centered on $v_x = 0$ m/s, so all we can know is that the atom's velocity is somewhere in the range $-4.3\times10^{-6}\text{ m/s} \le v_x \le 4.3\times10^{-6}$ m/s and thus its speed is in the range $0\text{ m/s} \le v \le 4.3\times10^{-6}$ m/s.

(b) With $v_{rms} = \frac{1}{2}u_{max} = 2.15\times10^{-6}\text{m/s}$, the lowest possible temperature is

$$T = \frac{mv_{rms}^2}{3k_B} = \frac{\left(23\times1.67\times10^{-27}\text{ kg}\right)\left(2.15\times10^{-6}\text{ m/s}\right)^2}{3\left(1.38\times10^{-23}\text{ J/K}\right)} = 4\times10^{-15}\text{ K}$$

Assess: The limit set on temperature by the uncertainty principle is much lower than 1 nK.

41

ONE-DIMENSIONAL QUANTUM MECHANICS

Exercises and Problems

41.1. Model: Model the electron as a particle in a rigid one-dimensional box of length L.
Solve: Absorption occurs from the ground state $n = 1$. It's reasonable to assume that the transition is from $n = 1$ to $n = 2$. The energy levels of an electron in a rigid box are

$$E_n = n^2 \frac{h^2}{8mL^2}$$

The absorbed photons must have just the right energy, so

$$E_{\text{ph}} = hf = \frac{hc}{\lambda} = \Delta E_{\text{elec}} = E_2 - E_1 = \frac{3h^2}{8mL^2}$$

$$\Rightarrow L = \sqrt{\frac{3h\lambda}{8mc}} = \sqrt{\frac{3\left(6.63\times10^{-34}\text{ J s}\right)\left(6.00\times10^{-7}\text{ m}\right)}{8\left(9.11\times10^{-31}\text{ kg}\right)\left(3.0\times10^{8}\text{ m/s}\right)}} = 7.39\times10^{-10}\text{ m} = 0.739\text{ nm}$$

41.3. Model: Model the electron as a particle in a rigid one-dimensional box of length L.
Solve: The energy levels for a particle in a rigid box are

$$E_n = n^2 \frac{\pi^2\hbar^2}{2L^2}$$

The wave function shown in Figure Ex41.3 corresponds to $n = 3$. This is also shown in Figure 41.7. Thus,

$$L = \frac{3\pi\hbar}{\sqrt{2mE_3}} = \frac{3h}{2\sqrt{2mE_3}} = \frac{3\left(6.63\times10^{-34}\text{ J s}\right)}{2\sqrt{2\left(9.11\times10^{-31}\text{ kg}\right)\left(6.0\text{ eV}\times1.6\times10^{-19}\text{ J/eV}\right)}} = 0.75\text{ nm}$$

41.7. Model: The wave function decreases exponentially in the classically forbidden region.
Solve: The probability of finding a particle in the small interval δx at position x is Prob(in δx at x) $= |\psi(x)|^2\,\delta x$. Thus the ratio

$$\frac{\text{Prob(in }\delta x\text{ at }x = L+\eta)}{\text{Prob(in }\delta x\text{ at }x = L)} = \frac{|\psi(L+\eta)|^2\,\delta x}{|\psi(L)|^2\,\delta x} = \frac{|\psi(L+\eta)|^2}{|\psi(L)|^2}$$

The wave function in the classically forbidden region $x \ge L$ is

$$\psi(x) = \psi_{\text{edge}} e^{-(x-L)/\eta}$$

At the edge of the forbidden region, at $x = L$, $\psi(L) = \psi_{\text{edge}}$. At $x = L + \eta$, $\psi(L + \eta) = \psi_{\text{edge}} e^{-1}$. Thus

$$\frac{\text{Prob(in }\delta x\text{ at }x = L+\eta)}{\text{Prob(in }\delta x\text{ at }x = L)} = \frac{|\psi(L+\eta)|^2}{|\psi(L)|^2} = \frac{(\psi_{\text{edge}} e^{-1})^2}{(\psi_{\text{edge}})^2} = e^{-2} = 0.135$$

41.11. Visualize:

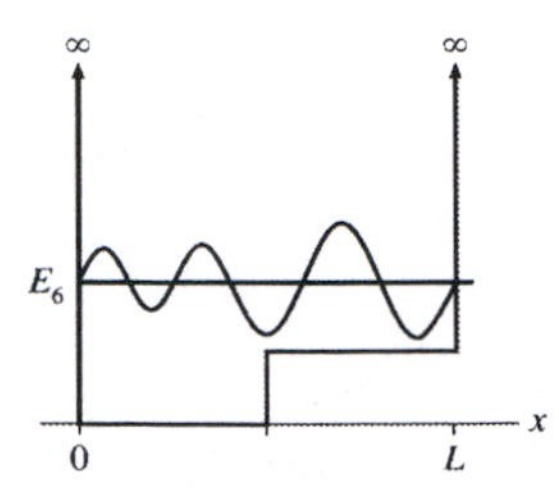

Solve: There are three factors to consider. First, the de Broglie wavelength increases as the particle's speed and kinetic energy decreases. Thus, the spacing between the nodes of $\psi(x)$ increases in regions where U is larger. Second, a particle is more likely to be found where it is moving the slowest. Thus, the amplitude of $\psi(x)$ increases in regions where U is larger. Third, for $n = 6$ there will be six antinodes to place.

41.19. Model: The electron is a quantum harmonic oscillator. The given levels are adjacent levels.
Solve: Let the two adjacent levels be n and $n + 1$. From Equation 41.48,

$$E_n = \left(n - \tfrac{1}{2}\right)\hbar\omega = 2.0 \text{ eV} \qquad E_{n+1} = \left(n + 1 - \tfrac{1}{2}\right)\hbar\omega = 2.8 \text{ eV}$$

$$\Rightarrow \frac{2.0 \text{ eV}}{n - \frac{1}{2}} = \frac{2.8 \text{ eV}}{n + \frac{1}{2}} \Rightarrow 2.0n + 1.0 = 2.8n - 1.4 \Rightarrow n = 3$$

Thus, $E_3 = \left(3 - \tfrac{1}{2}\right)\hbar\omega = 2.5\hbar\omega = 2.0 \text{ eV}$. Using Equation 41.43 for the angular frequency,

$$2.5\hbar\sqrt{\frac{k}{m}} = 2.0 \text{ eV} \Rightarrow k = m\left(\frac{2.0 \text{ eV}}{2.5\hbar}\right)^2 = \left(9.11\times10^{-31} \text{ kg}\right)\left[\frac{(2.0 \text{ eV})\left(1.6\times10^{-19} \text{ J/eV}\right)}{(2.5)\left(1.05\times10^{-34} \text{ J s}\right)}\right]^2 = 1.4 \text{ N/m}$$

41.25. Model: Model the particle as a particle in a rigid one-dimensional box of length L.
Solve: **(a)** From Equation 41.22, the particle's energies are

$$E_n = \frac{n^2h^2}{8mL^2} \Rightarrow E_2 - E_1 = \frac{h^2}{8mL^2}\left(2^2 - 1^2\right) = \frac{3h^2}{8mL^2}$$

Since $E_2 - E_1 = hf = hc/\lambda_{2\to1}$, we have $\lambda_{2\to1} = 8mcL^2/3h$.
(b) The length of the box is

$$L = \sqrt{\frac{3h\lambda_{2\to1}}{8mc}} = \sqrt{\frac{3\left(6.63\times10^{-34} \text{ J s}\right)\left(694\times10^{-9} \text{ m}\right)}{8\left(9.11\times10^{-31} \text{ kg}\right)\left(3.0\times10^{8} \text{ m/s}\right)}} = 0.795 \text{ nm}$$

41.29. Model: Model the particle as being confined in a rigid one-dimensional box of length L.
Visualize:

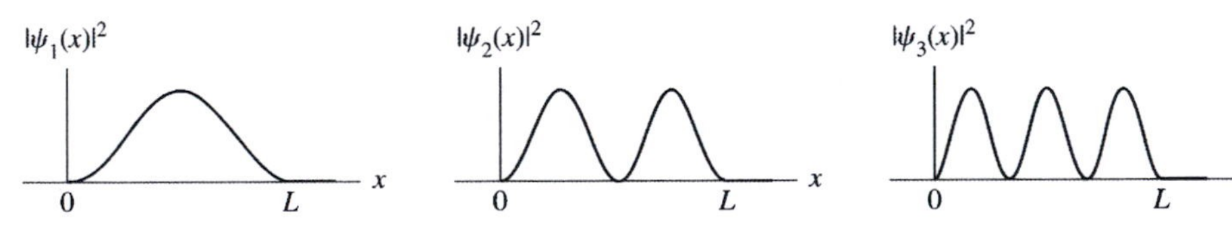

Solve: **(a)** The probability density is $|\psi_n(x)|^2 = (2/L)\sin^2(n\pi x/L)$. Graphs of $|\psi_1(x)|^2$, $|\psi_2(x)|^2$, and $|\psi_2(x)|^2$ are shown above.

(b) The particle is most likely to be found at x where $|\psi(x)|^2$ is a maximum. See table in part (d).

(c) The particle is least likely to be found at x where $|\psi(x)|^2 = 0$. See table in part (d).

(d) The probability of finding the particle in the left one-third of the box is the area under the $|\psi(x)|^2$ curve between $x = 0$ and $x = \frac{1}{3}L$. From examining the graphs, we can determine whether this is more than, less than, or equal to one-third of the total area. The results are shown in the table below.

n	Most likely	Least likely	Probability in left one-third
1	$\frac{1}{2}L$	0 and L	$<\frac{1}{3}$
2	$\frac{1}{4}L$ and $\frac{3}{4}L$	0, $\frac{1}{2}L$, and L	$>\frac{1}{3}$
3	$\frac{1}{6}L$, $\frac{3}{6}L$, and $\frac{5}{6}L$	0, $\frac{1}{3}L$, $\frac{2}{3}L$, and L	$=\frac{1}{3}$

(e) The probability of finding the particle in the range $0 \le x \le \frac{1}{3}L$ is

$$\text{Prob}\left(0 \le x \le \tfrac{1}{3}L\right) = \int_0^{L/3} |\psi_n(x)|^2\, dx = \frac{2}{L}\int_0^{L/3} \sin^2\left(\frac{n\pi x}{L}\right) dx$$

Change the variable to $u = n\pi x/L$. Then, $dx = (L/n\pi)du$. The integration limits become $u = 0$ at $x = 0$ m and $u = n\pi/3$ at $x = \frac{1}{3}L$. Then,

$$\text{Prob}\left(0 \le x \le \tfrac{1}{3}L\right) = \frac{2}{n\pi}\int_0^{n\pi/3} \sin^2 u\,du = \frac{2}{n\pi}\left[\tfrac{1}{2}u - \tfrac{1}{4}\sin 2u\right]_0^{n\pi/3} = \frac{1}{3} - \frac{1}{2n\pi}\sin\left(\frac{2n\pi}{3}\right)$$

The probability is 0.195 for $n = 1$, 0.402 for $n = 2$, and 0.333 for $n = 3$.

Assess: The results agree with the earlier estimates of the probability.

41.31. Model: The nucleus can be modeled as a potential well.

Visualize:

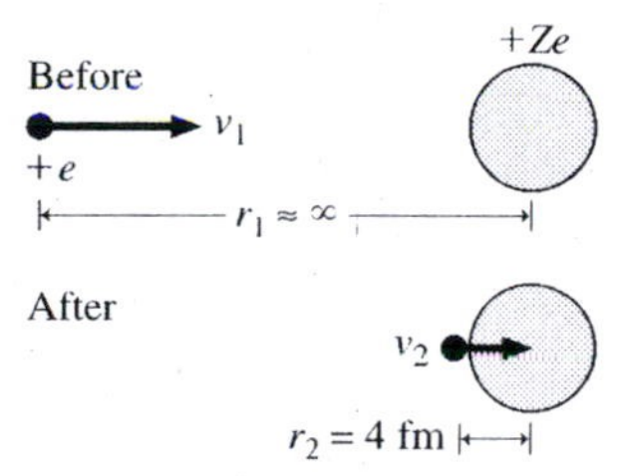

Please refer to Figure 41.17.

Solve: The gamma ray wavelength $\lambda = 1.73 \times 10^{-4}$ nm corresponds to a photon energy of $E_{\text{photon}} = hc/\lambda = 7.2$ MeV. From Fig. 41.17, we can see that a photon of this energy is emitted in a transition from the $n = 2$ to $n = 1$ energy level. This can happen after a proton-nucleus collision if the proton's impact excites the nucleus from the $n = 1$ ground state to the $n = 2$ excited state. To cause such an excitation, the proton's kinetic energy at the instant of impact must be $K \ge 7.2$ MeV. Let v_1 be the proton's initial speed at the distance $r_1 \approx \infty$. If v_1 is the *minimum* speed that can excite the $n = 2$ state in the nucleus, then the proton has $K_2 = 7.2$ MeV at the distance r_2 equal to the radius of nucleus (4 fm). Its potential energy at this point is the electrostatic potential energy between the proton of charge $+e$ and the nucleus of charge $+Ze$, with $Z = 13$. The conservation of energy equation $K_1 + U_1 = K_2 + U_2$ is

$$\tfrac{1}{2}mv_1^2 + 0\text{ J} = K_2 + \frac{Ze^2}{4\pi\varepsilon_0 r}$$

$$\Rightarrow v_1 = \sqrt{\frac{2}{m}\left(K_2 + \frac{Ze^2}{4\pi\varepsilon_0 r}\right)} = \sqrt{\frac{2}{1.67\times10^{-27}\text{ kg}}\left((7.2\text{ MeV})\left(\frac{1.60\times10^{-19}\text{ J}}{1\text{ eV}}\right) + \frac{(9.0\times10^9\text{ Nm}^2/\text{C}^2)13(1.6\times10^{-19}\text{ C})^2}{4.0\times10^{-15}\text{ m}}\right)}$$

$$= 4.77\times10^7\text{ m/s}$$

This is the *minimum* speed, so any $v_1 \ge 4.77 \times 10^7$ m/s can cause the emission of a gamma ray.

41.35. Solve: **(a)** The ground-state wave function of the quantum harmonic oscillator is $\psi_1(x) = A_1 e^{-x^2/2b^2}$. Normalization requires

$$\int_{-\infty}^{\infty} |\psi_1(x)|^2 dx = A_1^2 \int_{-\infty}^{\infty} e^{-x^2/2b^2} dx = 1$$

Change the variable to $u = x/b$. Then, $dx = bdu$. The integration limits don't change, so

$$1 = bA_1^2 \int_{-\infty}^{\infty} e^{-u^2} du$$

The definite integral can be looked up in a table of integrals. The result is $\sqrt{\pi}$. Hence,

$$1 = bA_1^2 \sqrt{\pi} = A_1^2 \sqrt{\pi b^2} \Rightarrow A_1 = \frac{1}{(\pi b^2)^{1/4}}$$

(b) The forbidden region is both $x < -b$ and $x > b$. $|\psi_1(x)|^2$ is symmetrical about $x = 0$ m, so

$$\text{Prob}(x < -b \text{ or } x > b) = (2)\text{Prob}(x > b) = 2\int_b^{\infty} |\psi_1(x)|^2 dx = \frac{2}{\sqrt{\pi b^2}} \int_b^{\infty} e^{-x^2/b^2} dx$$

(c) The integral of part (b) cannot be evaluated in closed form, but the answer can be found with a numerical integration. First, change the variable to $u = x/b$, making $dx = bdu$. But unlike the variable change in part (c), this *does* change the lower limit of integration. Thus,

$$\text{Prob}(x < -b \text{ or } x > b) = \frac{2}{\sqrt{\pi b^2}} b \int_1^{\infty} e^{-u^2} du = \frac{2}{\sqrt{\pi}} \int_1^{\infty} e^{-u^2} du$$

The definite integral can be evaluated numerically with a calculator or computer, giving

$$\int_1^{\infty} e^{-u^2} du = 0.139$$

The probability of finding the harmonic oscillator in the forbidden region is $2(\pi)^{-\frac{1}{2}}(0.139) = 0.157 = 15.7\%$.

41.37. Model: The collisions with the ground are perfectly elastic.
Solve: **(a)** The classical probability density at position y of finding a ball that bounces between the ground and height h is given by Equation 41.32:

$$P_{\text{class}}(y) = \frac{2}{Tv(y)}$$

where $v(y)$ is the ball's velocity as a function of y and T is the period of oscillation. For a freely falling object, energy conservation gives

$$mgh = \tfrac{1}{2}mv^2 + mgy \Rightarrow v(y) = \sqrt{2g(h-y)}$$

The time $t = \frac{1}{2}T$ to reach a height h after a collision with the ground can be found from kinematics:

$$\Delta y = h = \tfrac{1}{2}gt^2 \Rightarrow t = \sqrt{\frac{2h}{g}}$$

$$\Rightarrow P_{\text{class}}(y) = \frac{2}{2\sqrt{2h/g}} \frac{1}{\sqrt{2g(h-y)}} = \sqrt{\frac{g}{2h}} \frac{1}{\sqrt{2g(h-y)}} = \frac{1}{2\sqrt{h}\sqrt{h-y}} = \left(\frac{1}{2h}\right) \frac{1}{\sqrt{1-(y/h)}}$$

(b)

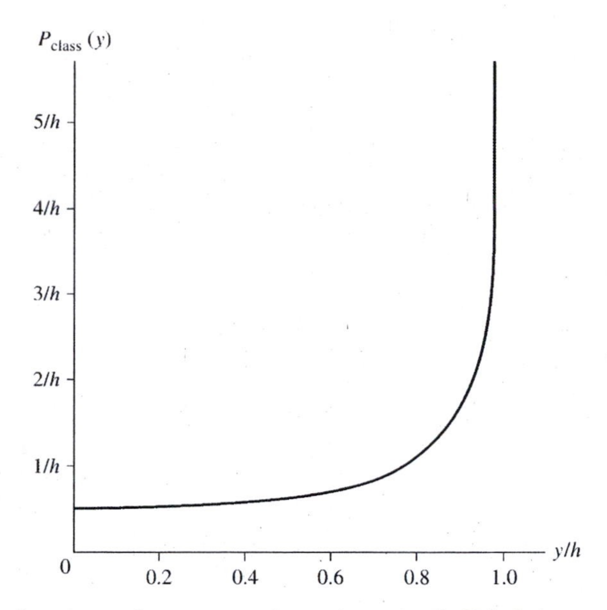

(c) The ball is most likely to be found near the upper turning point at $y = h$. This is because $v \to 0$ m/s at $y = h$ so the ball spends more time at this point. For the same reason, the ball spends the least time near the ground, where it is moving fastest, and thus the probability density is the least at $y = 0$ m.

42

ATOMIC PHYSICS

Exercises and Problems

42.3. Solve: **(a)** The orbital angular momentum is $L = \sqrt{l(l+1)}\hbar$. Thus,

$$l(l+1) = \left(\frac{L}{\hbar}\right)^2 = \left(\frac{3.65 \times 10^{-34}\ \text{J s}}{1.05 \times 10^{-34}\ \text{J s}}\right)^2 = 12 \Rightarrow l = 3$$

This is an f electron.
(b) The l quantum number is required to be less than n. Thus, the minimum possible value of n for an electron in the f state is $n_{\text{min}} = 4$. The corresponding minimum possible energy is

$$E_{\text{min}} = E_4 = -\frac{13.60\ \text{eV}}{4^2} = -0.85\ \text{eV}$$

42.7. Solve: **(a)** A lithium atom has three electrons, two are in the 1s shell and one is in the 2s shell. The electron in the 2s shell has the following quantum numbers: $n = 2$, $l = 0$, $m = 0$, and m_s. m_s could be either $+\frac{1}{2}$ or $-\frac{1}{2}$. Thus, lithium atoms should behave like hydrogen atoms because lithium atoms could exist in the following two states: $\left(2, 0, 0, +\frac{1}{2}\right)$ and $\left(2, 0, 0, -\frac{1}{2}\right)$. Thus there are two lines.

(b) For a beryllium atom, we have two electrons in the 1s shell and two electrons in the 2s shell. The electrons in both the 1s and 2s states are filled. Because the two electron magnetic moments point in opposite directions, beryllium has *no* net magnetic moment and is not deflected in a Stern-Gerlach experiment. Thus there is only one line.

42.11. Solve: **(a)** Ten electrons (Z = 10) make the element neon (Ne). These are *not* the ten lowest energy states because $1s^2 2s^2 2p^6$ would be lower in energy than $1s^2 2s^2 2p^5 3d$. This is an excited state of Ne.
(b) Twenty-six electrons (Z = 26) make the element iron (Fe). These are *not* the 26 lowest energy states because the 3d shell is not filled. This is the ground state of Fe.

42.15. Solve: **(a)** A $4p \rightarrow 4s$ transition is allowed because $\Delta l = 1$. Using the sodium energy levels from Figure 42.25, the wavelength is

$$\lambda = \frac{hc}{\Delta E} = \frac{1240\ \text{eV nm}}{3.75\ \text{eV} - 3.19\ \text{eV}} = 2210\ \text{nm} = 2.21\ \mu\text{m}$$

(b) A $3d \rightarrow 4s$ transition is not allowed because $\Delta l = 2$ violates the selection rule that requires $\Delta l = 1$.

42.17. Solve: The interval $\Delta t = 0.50$ ns is very small in comparison with the lifetime $\tau = 25$ ns, so we can write

$$\text{Prob(decay in } \Delta t) = r\Delta t$$

where r is the decay rate. The decay rate is related to the lifetime by $r = 1/\tau = 1/(25\ \text{ns}) = 0.040\ \text{ns}^{-1}$. Thus

$$\text{Prob(decay in } \Delta t) = (0.040\ \text{ns}^{-1})(0.50\ \text{ns}) = 0.020 = 2.0\%$$

42.23. Solve: **(a)** The wavelength is

$$\lambda = \frac{hc}{\Delta E} = \frac{1240\ \text{eV nm}}{1.17\ \text{eV} - 0\ \text{eV}} = 1060\ \text{nm} = 1.06\ \mu\text{m}$$

(b) The energy per photon is E_{ph} = 1.17 eV = 1.87×10^{-19} J. The power output of the laser is the number of photons per second times the energy per photon:

$$P = (1.0 \times 10^{19}\ \text{s}^{-1})(1.87 \times 10^{-19}\ \text{J}) = 1.9\ \text{J/s} = 1.9\ \text{W}$$

42.25. Solve: **(a)** For $s = 1$, $S = \sqrt{s(s+1)}\hbar = \sqrt{2}\hbar$

(b) The spin quantum number is $m_s = -1$, 0, or 1.

(c) The figure below shows the three possible orientations of $\vec{S}$.

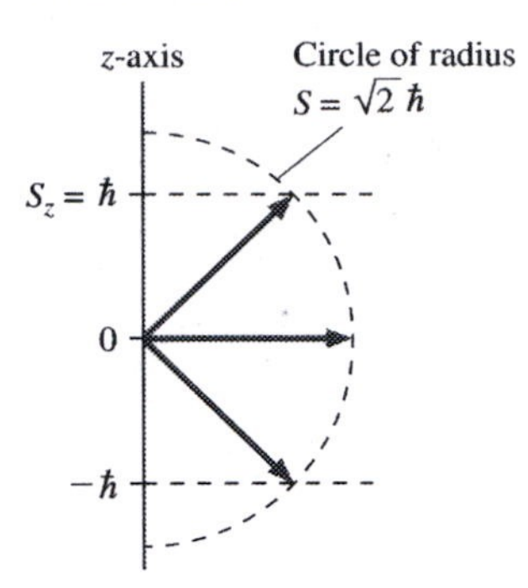

42.29. Solve: **(a)** From Equations 42.7 and 42.20, the radial wave function and radial probability density are

$$R_{1s}\left(\tfrac{1}{2}a_B\right) = \frac{1}{\sqrt{\pi a_B^3}}e^{-\frac{1}{2}} = \frac{0.607}{\sqrt{\pi a_B^3}} \Rightarrow P_r\left(\tfrac{1}{2}a_B\right) = 4\pi\left(\frac{a_B}{2}\right)^2\left(\frac{0.607}{\sqrt{\pi a_B^3}}\right)^2 = \frac{0.368}{a_B}$$

From Equation 42.9, the probability is

$$\text{Prob(in } \delta r \text{ at } r) = P_r(r)\delta r = P_r\left(\tfrac{1}{2}a_B\right)\left[(0.01)a_B\right] = \left(\frac{0.368}{a_B}\right)(0.01a_B) = 3.7 \times 10^{-3}$$

(b) Likewise,

$$R_{1s}(a_B) = \frac{1}{\sqrt{\pi a_B^3}}e^{-1} = \frac{0.368}{\sqrt{\pi a_B^3}} \Rightarrow P_r(a_B) = 4\pi a_B^2\frac{(0.368)^2}{\pi a_B^3} = \frac{0.541}{a_B}$$

The probability is $P_r(r)\delta r = P_r(a_B)(0.01)a_B = 5.4 \times 10^{-3}$.

(c) For $r = 2a_B$,

$$R_{1s}(2a_B) = \frac{1}{\sqrt{\pi a_B^3}}e^{-2} = \frac{0.135}{\sqrt{\pi a_B^3}} \Rightarrow P_r(2a_B) = 4\pi(2a_B)^2\frac{(0.135)^2}{\pi a_B^3} = \frac{0.293}{a_B}$$

The probability is $P_r(r)\delta r = P_r(2a_B)(0.01)a_B = 2.9 \times 10^{-3}$.

42.31. Solve: From Equation 42.7, the $2p$ radial wave function of the hydrogen atom can be written

$$R_{2p}(r) = A_{2p}\left(\frac{r}{2a_B}\right)e^{-r/2a_B}$$

The normalization condition for the three-dimensional hydrogen atom is

$$\int_0^\infty P_r(r)dr = 4\pi\int_0^\infty r^2\left|R_{nl}(r)\right|^2 dr = 4\pi\int_0^\infty A_{2p}^2\left(\frac{r^2}{4a_B^2}\right)e^{-\frac{2r}{2a_B}}r^2 dr$$

$$= \pi\frac{A_{2p}^2}{a_B^2}\int_0^\infty r^4 e^{-r/a_B}dr = \frac{A_{2p}^2\pi}{a_B^2}\left[\frac{4!}{(1/a_B)^5}\right] = 24\pi a_B^3 A_{2p}^2 = 1 \Rightarrow A_{2p} = \sqrt{\frac{1}{24\pi a_B^3}}$$

42.39. Visualize: Please refer to Figure 42.25.
Solve: A photon wavelength of 818 nm corresponds to an energy of

$$E = \frac{hc}{\lambda} = \frac{1240 \text{ eV nm}}{818 \text{ nm}} = 1.516 \text{ eV}$$

From Figure 42.25, the transition that obeys the selection rule $\Delta l = 1$ and has a magnitude around 1.5 eV is $3p \rightarrow 3d$. Note that $E_{3d} - E_{3p} = 3.620 \text{ eV} - 2.104 \text{ eV} = 1.516 \text{ eV}$, which is exactly equal to the photon energy. The atom was excited to the $3d$ state from the ground state. Thus the minimum kinetic energy of the electron was

$$\frac{1}{2}mv^2 = 3.62 \text{ eV} \Rightarrow v = \sqrt{\frac{2(3.62 \text{ eV})(1.60\times10^{-19} \text{ J/eV})}{9.11\times10^{-31} \text{ kg}}} = 1.13\times10^6 \text{ m/s}$$

42.43. Model: We have a one-dimensional rigid box with infinite potential walls and a length 0.50 nm.
Solve: **(a)** From Equation 41.22, the lowest energy level is

$$E_1 = \frac{h^2}{8mL^2} = \frac{(6.63\times10^{-34} \text{ J s})^2}{8(9.11\times10^{-31} \text{ kg})(5.0\times10^{-10} \text{ m})^2}\frac{1 \text{ eV}}{1.60\times10^{-19} \text{ J}} = 1.51 \text{ V}$$

The next two levels are $E_2 = 4E_1 = 6.04$ eV and $E_3 = 9E_1 = 13.6$ eV. The Pauli principle allows only two electrons in each of these energy levels, one with spin up and one with spin down. So five electrons fill the $n = 1$ and $n = 2$ levels, with the fifth electron going to $n = 3$.

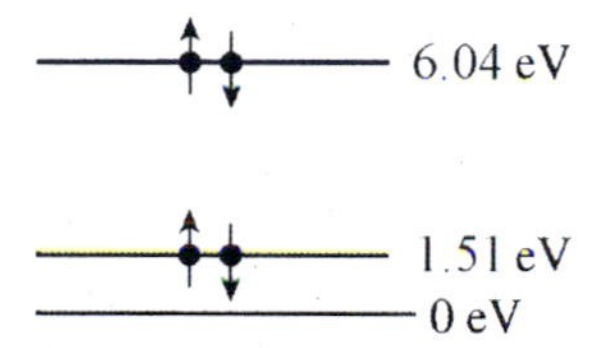

(b) The ground-state energy of these five electrons is $E = 2E_1 + 2E_2 + E_3 = 28.7$ eV

42.45. Solve: **(a)** From Table 42.3, the lifetime of the $2p$ state of hydrogen is $\tau = 1.6$ ns. The decay rate is

$$r = \frac{1}{\tau} = \frac{1}{1.6\times10^{-9} \text{ s}} = 6.25\times10^8 \text{ s}^{-1}$$

(b) From Equation 42.25, the number of excited atoms left at time t is $N_{exc} = N_0 e^{-t/\tau}$. If 10% of a sample decays, then 90% of the atoms in the sample are still excited. That is, $N_{exc} = 0.90N_0$. The time for this to occur is calculated as follows:

$$N_{exc} = 0.90N_0 = N_0 e^{-t/\tau} \Rightarrow e^{-t/\tau} = 0.90 \Rightarrow t = -\tau \ln 0.90 = 0.17 \text{ ns}$$

42.47. Solve: The number of excited atoms left at time t is given by Equation 42.25: $N_{ext} = N_0 e^{-t/\tau}$. If 1% of the atoms in the excited state decay in $t = 0.20$ ns, then 99% of the atoms remain in the excited state. So,

$$0.99N_0 = N_0 e^{-0.20 \text{ ns}/\tau} \Rightarrow 0.99 = e^{-0.20 \text{ ns}/\tau} \Rightarrow \ln(0.99) = -0.20 \text{ ns}/\tau \Rightarrow \tau = 19.90 \text{ ns}$$

Having determined τ, we can now find the time during which 25% of the sample of excited atoms would decay, leaving 75% still excited. Applying Equation 42.25 once again,

$$0.75N_0 = N_0 e^{-t/19.90 \text{ ns}} \Rightarrow t = 5.7 \text{ ns}$$

43

NUCLEAR PHYSICS

Exercises and Problems

43.3. Solve: **(a)** The radius and diameter of the nucleus of ^{4}He are

$$r = r_0 A^{1/3} = (1.2 \text{ fm})(4)^{1/3} = 1.90 \text{ fm} \Rightarrow d = 3.8 \text{ fm}$$

(b) For ^{40}Ar, $r = (1.2 \text{ fm})(40)^{1/3} = 4.10$ fm and $d = 8.2$ fm.
(c) For ^{220}Rn, $r = (1.2 \text{ fm})(220)^{1/3} = 7.24$ fm and $d = 14.5$ fm.

43.7. Solve: The nuclear density was found to be 2.3×10^{17} kg/m^3. Thus

$$M = \rho_{\text{nuclear}} V = \left(2.3\times10^{17} \text{ kg/m}^3\right)\left(\frac{4\pi}{3}\right)\left(0.5\times10^{-2} \text{ m}\right)^3 = 1.2\times10^{11} \text{ kg}$$

Assess: The nuclear density is tremendously large compared to the density of familiar liquids and solids.

43.11. Solve: From Equation 43.6, the binding energy for ^{54}Fe is

$$B = Zm_{\text{H}} + Nm_{\text{n}} - m_{\text{atom}} = 26(1.00783 \text{ u}) + 32(1.00866 \text{ u}) - 57.933278 \text{ u}$$
$$= 0.54742 \text{ u} \times 931.49 \text{ MeV/u} = 510 \text{ MeV}$$

The binding energy per nucleon is $\frac{1}{58}(509.92 \text{ MeV}) = 8.79$ MeV.

For ^{58}Ni, the binding energy is

$$B = Zm_{\text{H}} + Nm_{\text{n}} - m_{\text{atom}} = 28(1.00783 \text{ u}) + 30(1.00866 \text{ u}) - 57.935346 \text{ u}$$
$$= 0.54369 \text{ u} \times 931.49 \text{ MeV/u} = 506 \text{ MeV}$$

The binding energy per nucleon is $\frac{1}{58}(506.45 \text{ MeV}) = 8.73$ MeV.

Assess: The binding energy per nucleon is higher for iron but the binding energy per nucleon for nickel isn't much different; we expect both of these results from Figure 43.6

43.13. Solve: From Equation 43.6, the binding energy for ^{12}C is

$$B = Zm_{\text{H}} + Nm_{\text{n}} - m_{\text{atom}} = 6(1.00783 \text{ u}) + 6(1.00866 \text{ u}) - 12.00000 \text{ u}$$
$$= 0.09894 \text{ u} \times 931.49 \text{ MeV/u} = 92.162 \text{ MeV}$$

The binding energy per nucleon is $\frac{1}{12}(92.162 \text{ MeV}) = 7.68$ MeV.

For ^{13}C, the binding energy is

$$B = Zm_{\text{H}} + Nm_{\text{n}} - m_{\text{atom}} = 6(1.00783 \text{ u}) + 7(1.00866 \text{ u}) - 13.00336 \text{ u}$$
$$= 0.10424 \text{ u} \times 931.49 \text{ MeV/u} = 97.099 \text{ MeV}$$

The binding energy per nucleon is $\frac{1}{13}(97.099 \text{ MeV}) = 7.47$ MeV. Thus, ^{12}C is slightly more tightly bound.

43.19. Solve: From Figure 43.8, the nuclear potential energy at $r = 1.0$ fm is

$$U_{\text{nuclear}} = -50 \text{ MeV} = -(50 \text{ MeV})(1.6 \times 10^{-19} \text{ J/eV}) = -8.0 \times 10^{-12} \text{ J}$$

The gravitational potential energy is

$$U_{\text{grav}} = -\frac{Gm^2}{r} = -\frac{(6.67\times10^{-11}\ \text{Nm}^2/\text{kg}^2)(1.67\times10^{-27}\ \text{kg})^2}{1.0\times10^{-15}\ \text{m}} = -1.86\times10^{-49}\ \text{J}$$

$$\Rightarrow \frac{U_{\text{grav}}}{U_{\text{nuclear}}} = \frac{1.86\times10^{-49}\ \text{J}}{8.0\times10^{-12}\ \text{J}} = 2.3\times10^{-38}$$

43.21. Solve: (a) The $A = 14$ nuclei listed in Appendix C are ^{14}C, ^{14}N, and ^{14}O. ^{14}C has $Z = 6$, so $N = 8$. ^{14}N has $Z = 7$, $N = 7$, and ^{14}O has $Z = 8$, $N = 6$. These 3 nuclei are the $A = 14$ isobars.

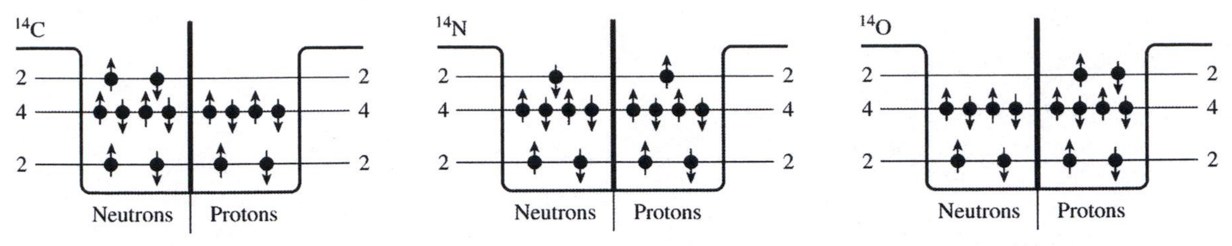

(b) ^{14}N is a stable nucleus, but ^{14}C and ^{14}O are radioactive. ^{14}C undergoes beta-minus decay and ^{14}O undergoes beta-plus decay. In beta-minus decay, a neutron within the nucleus changes into a proton and an electron. In beta-plus decay, a proton changes into a neutron and a positron.

Assess: The above decays for ^{14}C and ^{14}O are consistent with the fact that the line of stability follows the $N = Z$ line for $Z < 16$.

43.27. Model: The activity R of a radioactive sample is the number of decays per second.

Solve: The rate of decay is

$$R = \left|\frac{dN}{dt}\right| = rN = \frac{N}{\tau} = \frac{(0.693)N}{t_{1/2}}$$

$$\Rightarrow t_{1/2} = \frac{(0.693)N}{R} = \frac{(0.693)(5.0\times10^{15}\ \text{atoms})}{5.0\times10^{8}\ \text{Bq}} = 6.93\times10^{6}\ \text{s}\times\frac{1\ \text{day}}{86{,}400\ \text{s}} = 80.2\ \text{days} \approx 80\ \text{days}$$

43.33. Model: The decay is $^{19}\text{O} \to {}^{19}\text{F} + e^- + \bar{\nu}$.

Solve: Beta-minus decay leaves the daughter atom as a positive ion. However, the mass of the ion plus the mass of the escaping electron are the mass of a neutral atom, which is what is tabulated in Appendix C. Thus the mass loss is the mass difference between the two neutral atoms. In Appendix C we find that $m(^{19}\text{O}) = 19.003577$ u and $m^{19}\text{F} = 18.998404$ u. The energy released in the beta-minus decay corresponds to a mass change of 0.005173 u. The energy released is

$$E = \Delta mc^2 = (0.005173\ \text{u})c^2 \times \frac{931.49\ \text{MeV}/c^2}{1\ \text{u}} = 4.82\ \text{MeV}$$

Assess: This energy is shared between the electron and the antineutrino.

43.37. Model: The radiation dose in units of rems is a combination of deposited energy and biological effectiveness. The RBE for beta radiation is 1.5.

Solve: 1 Gy is defined as 1.00 J/kg of absorbed energy. In the case of the 50 kg worker, the energy absorbed per kg is

$$\frac{20\times10^{-3}\ \text{J}}{50\ \text{kg}} = 4.0\times10^{-4}\ \text{J/kg}$$

This corresponds to a dose of

$$4.0\times10^{-4}\ \text{J/kg}\times\frac{1\ \text{Gy}}{1.00\ \text{J/kg}} = 4.0\times10^{-4}\ \text{Gy}$$

The dose equivalent in rems is $(4.0\times10^{-4}\ \text{Gy})(1.5) = 0.060$ Sv = 60 mrem.

43.39. Visualize: The radii of the alpha particle and the gold nucleus are

$$r_\alpha = (1.2\text{ fm})(4)^{1/3} = 1.90\text{ fm} \qquad r_{\text{Au}} = (1.2\text{ fm})(197)^{1/3} = 6.98\text{ fm}$$

If the alpha just touches the surface of the gold nucleus, the distance between their centers is $r_\text{f} = 1.90\text{ fm} + 6.98\text{ fm} = 8.88\text{ fm}$.

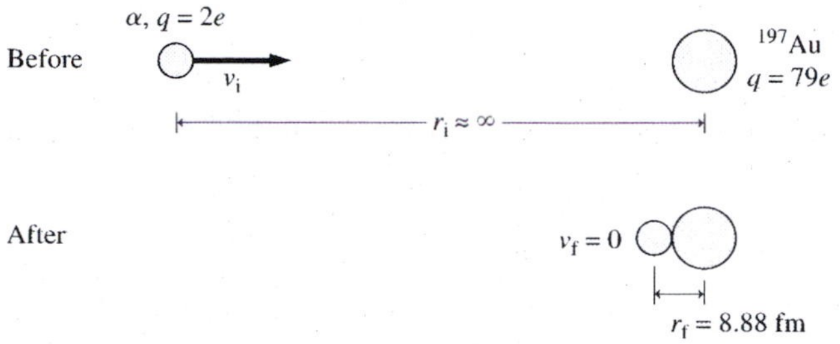

Before α, $q = 2e$ v_i ^{197}Au $q = 79e$ $r_\text{i} \approx \infty$

After $v_\text{f} = 0$ $r_\text{f} = 8.88$ fm

Solve: **(a)** The energy conservation equation $K_\text{f} + U_\text{f} = K_\text{i} + U_\text{i}$ is

$$0\text{ J} + \frac{1}{4\pi\varepsilon_0}\frac{(2e)(79e)}{8.88\times10^{-15}\text{ fm}} = \tfrac{1}{2}mv_\text{i}^2 + 0\text{ J}$$

$$\Rightarrow v_\text{i} = \sqrt{\frac{2(9.0\times10^9\text{ N m}^2/\text{C}^2)(158)(1.60\times10^{-19}\text{ C})^2}{(4\times1.661\times10^{-27}\text{ kg})(8.88\times10^{-15}\text{ fm})}} = 3.51\times10^7\text{ m/s}$$

(b) The energy of the alpha particle is

$$K = \tfrac{1}{2}mv_\text{i}^2 = \tfrac{1}{2}(4\times1.661\times10^{-27}\text{ kg})(3.51\times10^7\text{ m/s})^2$$

$$= 4.09\times10^{-12}\text{ J}\times(1\text{ MeV}/1.60\times10^{-13}\text{ J}) = 25.6\text{ MeV}$$

43.41. Solve: **(a)** The sun's mass of 1.99×10^{30} kg is unchanged, but it assumes the density of nuclear matter, which we found to be $\rho_\text{nuc} = 2.3\times10^{17}\text{ kg/m}^3$. The volume of the collapsed sun is

$$V = \frac{4}{3}\pi r^3 = \frac{M_\text{S}}{\rho_\text{nuc}} = \frac{1.99\times10^{30}\text{ kg}}{2.3\times10^{17}\text{ kg/m}^3} = 8.65\times10^{12}\text{ m}^3$$

Thus its radius is

$$r = \left(\frac{3(8.65\times10^{12}\text{ m}^3)}{4\pi}\right)^{1/3} = 12{,}700\text{ m} = 12.7\text{ km}$$

(b) We can use the conservation of angular momentum to find the new rotational period:

$$(I\omega)_\text{after} = (I\omega)_\text{before} \Rightarrow \omega_\text{after} = \frac{I_\text{before}}{I_\text{after}}\omega_\text{before} = \frac{\frac{2}{5}M_\text{S}R_\text{S}^2}{\frac{2}{5}M_\text{S}r^2}\omega_\text{before}$$

$$\Rightarrow \frac{2\pi}{T_\text{after}} = \left(\frac{R_\text{S}}{r}\right)^2\frac{2\pi}{T_\text{before}} \Rightarrow T_\text{after} = \left(\frac{1.27\times10^4\text{ m}}{6.96\times10^8\text{ m}}\right)^2(27\text{ days}) = 7.8\times10^{-4}\text{ s} = 780\ \mu\text{s}$$

43.43. Solve: **(a)** The binding energy of the electron in a hydrogen atom is $B = 13.6$ eV. That is, the mass decreases by the equivalent of 13.6 eV when an electron and proton form a hydrogen atom. Since $B = \Delta mc^2$,

$$\Delta m = \frac{13.6\text{ eV}}{c^2} = \frac{13.6\text{ eV}}{c^2}\times\frac{1\text{ u}}{931.49\text{ MeV}/c^2} = 1.46\times10^{-8}\text{ u}$$

As a percentage of the hydrogen mass, the mass decrease is

$$\frac{\Delta m}{1.007825\text{ u}} = \frac{1.46\times10^{-8}\text{ u}}{1.007825\text{ u}} = 1.45\times10^{-6}\%$$

(b) The mass decrease is $\Delta m = 2m_\text{p} + 2m_\text{n} - m_\text{He nuc}$. These are nuclear masses, but Appendix C tabulates atomic masses. Add and subtract the mass of two electrons:

$$\Delta m = 2(m_\text{p} + m_\text{e}) + 2m_\text{n} - (m_\text{He nuc} - 2m_\text{e}) = 2m_\text{H} + 2m_\text{n} - m_\text{He}$$

where m_H is the mass of a hydrogen atom and m_{He} is the mass of a helium atom. Using Appendix C,

$$\Delta m = 2(1.007825 \text{ u}) + 2(1.008665) - 4.002602 = 0.0304 \text{ u}$$

As a percentage of the helium mass, the mass decrease is

$$\frac{0.0304 \text{ u}}{4.0026 \text{ u}} = 0.0076 = 0.76\%$$

(c) Although mass does change in chemical reactions, the change is an incredibly small fraction of the mass of the atoms. No experiment will be sensitive to changes of $\approx 1 \times 10^{-6}\%$, so this small change in mass is easily neglected. Not so in nuclear reactions, where the mass change can be $\approx 1\%$ of the particle masses. Not only is this mass change easily detectable, it is essential for understanding nuclear reactions.

43.51. Solve: From Appendix C, the half-life of ^{3}H is $t_{1/2} = 12.33$ years. The activity of a radioactive sample is defined as $R = rN$. The decay rate is

$$r = \frac{1}{\tau} = \frac{0.693}{t_{1/2}} = \frac{0.693}{12.33 \text{ years}} \times \frac{1 \text{ year}}{3.15 \times 10^7 \text{ s}} = 1.784 \times 10^{-9} \text{ s}^{-1}$$

The number of atoms in the sample is

$$N = \frac{2.0 \times 10^{-3} \text{g}}{3 \text{ g/mol}} \times 6.02 \times 10^{23} \text{ mol}^{-1} = 4.01 \times 10^{20}$$

$$\Rightarrow R = rN = (1.784 \times 10^{-9} \text{ s}^{-1})(4.01 \times 10^{20}) = 7.16 \times 10^{11} \text{ Bq} = 7.16 \times 10^{11} \text{ Bq} \times \frac{1 \text{ Ci}}{3.7 \times 10^{10} \text{ Bq}} = 19.4 \text{ Ci}$$

43.53. Solve: The half-life of a ^{133}Cs sample is

$$t_{1/2} = 30 \text{ years} \times \frac{3.15 \times 10^7 \text{ s}}{1 \text{ year}} = 9.45 \times 10^8 \text{ s}$$

The activity of a radioactive sample is $R = rN$. The decay rate is

$$r = \frac{1}{\tau} = \frac{0.693}{t_{1/2}} = \frac{0.693}{9.45 \times 10^8 \text{ s}} = 7.33 \times 10^{-10} \text{ s}^{-1}$$

$$\Rightarrow N = \frac{R_0}{r} = \frac{2.0 \times 10^8 \text{ Bq}}{7.33 \times 10^{-10} \text{ s}^{-1}} = 2.73 \times 10^{17} \text{ atoms}$$

That is, at $t = 0$ s, we have 2.73×10^{17} ^{137}Cs atoms and after a very long time all will have decayed. Thus 2.73×10^{17} beta particles will have been emitted.

43.55. Model: The decay is $^{223}\text{Ra} \rightarrow {}^{219}\text{Rn} + {}^4\text{He}$.

Solve: (a) The energy released in the above decay is

$$E = \left[m\left({}^{223}\text{Ra}\right) - m\left({}^{219}\text{Rn}\right) - m\left({}^4\text{He}\right)\right] \times 931.49 \text{ MeV/u}$$

$$= \left[223.018499 \text{ u} - 219.009477 \text{ u} - 4.002602 \text{ u}\right] \times 931.49 \text{ MeV/u} = 5.98 \text{ MeV}$$

That is, each α-particle is released with an energy of

$$5.98 \text{ MeV} = 5.98 \times 10^6 \text{ eV} \times 1.6 \times 10^{-19} \text{ J/eV} = 9.57 \times 10^{-13} \text{ J}$$

The amount of energy needed to raise the temperature of 100 mL of water at 18°C to 100°C is

$$Q = mc\Delta t = (0.10 \text{ L})\left(\frac{1 \text{ kg}}{1 \text{ L}}\right)(4190 \text{ J/kg K})(100 \text{ K} - 18 \text{ K}) = 34{,}400 \text{ J}$$

The number of decays we need to generate this amount of energy is

$$\frac{34{,}400 \text{ J}}{9.57 \times 10^{-13} \text{ J}} = 3.59 \times 10^{16}$$

The total number of radium atoms in the cube at $t = 0$ s is

$$N_0 = \frac{1\text{ g}}{223\text{ g/mol}} \times 6.02\times10^{23}\text{ atoms/mol} = 2.70\times10^{21}\text{ atoms}$$

The number of needed decays is very small compared to N_0, so we can write

$$\frac{dN}{dt} \approx \frac{\Delta N}{\Delta t} = -rN \Rightarrow \Delta t = -\frac{\Delta N}{rN} = -\tau\frac{\Delta N}{N} = -\frac{t_{1/2}}{\ln 2}\frac{\Delta N}{N}$$

where we used $r = 1/\tau$. $\Delta N = -3.59 \times 10^{16}$, with the negative sign due to the fact that the number of radium atoms decreases. Since $t_{1/2} = 11.43$ days $= 9.876 \times 10^5$ s, the time for 3.59×10^{16} decays is

$$\Delta t = -\frac{9.876\times10^5\text{ s}}{\ln 2}\frac{(-3.59\times10^{16})}{2.70\times10^{21}} = 18.9\text{ s} \approx 19\text{ s}$$

(b) It's possible that an alpha-particle collision will break the molecular bond in a *very* small number of H_2O molecules, causing H_2 gas and O_2 gas to bubble out of the water. The water that remains in the container has not changed or been altered.

43.61. Model: Assume that the tracer stays in the body for the life of the tracer.
Solve: The initial activity of the source is

$$R_0 = 30\ \mu\text{Ci} = 30\times10^{-6}\text{ Ci} \times \frac{3.7\times10^{10}\text{ Bq}}{1\text{ Ci}} = 1.11\times10^6\text{ Bq}$$

The decay rate is

$$r = \frac{1}{\tau} = \frac{\ln 2}{t_{1/2}} = \frac{\ln 2}{5.0\text{ d}\times 24\text{ h/d}\times 3600\text{ s/h}} = 1.605\times10^{-6}\text{ s}^{-1}$$

The number of atoms that will decay by beta emission is

$$N_0 = \frac{R_0}{r} = \frac{1.11\times10^6\text{ Bq}}{1.605\times10^{-6}\text{ s}^{-1}} = 6.916\times10^{11}$$

Since 10% of the beta particles escape, there are 0.90 N_0 disintegrations in the body. Since the average energy of the beta particle is 0.35 MeV, the energy absorbed by the body is

$$E = 0.35\text{ MeV} \times (1.6 \times 10^{-19}\text{ J/eV}) \times 0.90 \times (6.926 \times 10^{11}) = 0.03485\text{ J}$$

Thus, the radiation dose is

$$\frac{0.03485\text{ J}}{75\text{ kg}} = 4.647\times10^{-4}\text{ J/kg} \times \frac{1\text{ Gy}}{1.00\text{ J/kg}} = 4.65\times10^{-4}\text{ Gy}$$

The radiation dose in rems is 4.65×10^{-4} Gy $\times$ 1.5 $= 6.97 \times 10^{-4}$ Sv $= 69.7$ mrem $\approx$ 70 mrem.

43.65. Model: The number of ^{239}Pu atoms decays exponentially.
Solve: **(a)** The number of ^{239}Pu atoms in a 1.0-μm-diameter particle is

$$N = \left(\frac{m}{M_A}\right)N_A = \frac{(\rho V)}{M_A}N_A = \frac{\rho(\frac{4\pi}{3})r^3 N_A}{M_A}$$

$$= \frac{(19{,}800\text{ kg/m}^3)(\frac{4\pi}{3})(0.5\times10^{-6}\text{ m})^3(6.02\times10^{23}\text{ mol}^{-1})}{239\times10^{-3}\text{ kg/mol}} = 2.61\times10^{10} \approx 2.6\times10^{10}$$

(b) The activity of the particle is

$$R = \left|\frac{dN}{dt}\right| = rN = \frac{\ln 2}{t_{1/2}}N = \frac{0.6931}{24{,}000\text{ years}\times 3.15\times10^7\text{ s/year}}\times 2.61\times10^{10} = 0.0239\text{ Bq} \approx 0.024\text{ Bq}$$

(c) The volume of the 50-μm diameter sphere of tissue around the particle is

$$\frac{4\pi}{3}(25\times10^{-6}\text{ m})^3 = 6.545\times10^{-14}\text{ m}^3$$

This volume of the tissue has a mass of

$$m = \rho V = \left(1000\ \text{kg/m}^3\right)\left(6.545\times10^{-14}\ \text{m}^3\right) = 6.545\times10^{-11}\ \text{kg}$$

In one year, the activity changes insignificantly. Thus the number of decays per year is

$$\left|\frac{dN}{dt}\right|\Delta t = \left(0.0239\ \text{Bq}\right)\left(3.15\times10^7\ \text{s}\right) = 7.529\times10^5$$

Since each decay creates an α-particle with energy 5.2 MeV, the total energy received per year by the tissue is

$$(7.529 \times 10^5 \times 5.2\ \text{MeV}) \times (1.6 \times 10^{-19}\ \text{J/eV}) = 6.264 \times 10^{-7}\ \text{J}$$

Dividing this energy by the tissue's mass, the dose received by the tissue is

$$6.264 \times 10^{-7}\ \text{J}/6.545 \times 10^{-11}\ \text{kg} = 9.57 \times 10^3\ \text{J/kg} = 9.57 \times 10^3\ \text{Gy}$$

The dose per year in rem is

$$(9.57 \times 10^3\ \text{Gy})(\text{RBE}) = (9.57 \times 10^3\ \text{Gy})(15) = 1.436 \times 10^5\ \text{Sv} = 1.436 \times 10^7\ \text{rem} \approx 1.4\times10^7\ \text{rem}$$

(d) This is a very high dose to a very small volume of body mass. Table 43.5 gives a typical exposure (in units of rem/year) from various radiation sources. The background radiation from various natural occurring sources is about 300 mrem. The exposure to this tissue is roughly 50 million times the background level.